Lecture Notes in Control and Information Sciences

Edited by A. V. Balakrishnan and M. Thoma

Vol. 1: Distributed Parameter Systems: Modelling and Identification
Proceedings of the IFIP Working Conference, Rome, Italy, June 21–26, 1976
Edited by A. Ruberti
V, 458 pages. 1978

Vol. 2: Rew Trends in Systems Analysis
International Symposium, Versailles, December 13–17, 1976
Edited by A. Bensoussan and J. L. Lions
VII, 759 pages. 1977

Vol. 3: Differential Games and Applications
Proceedings of a Workshop, Enschede, Netherlands, March 16–25, 1977
Edited by P. Hagedorn, H. W. Knobloch, and G. J. Olsder
XII, 236 pages. 1977

Vol. 4: M. A. Crane, A. J. Lemoine
An Introduction to the Regenerative Method for Simulation Analysis
VII, 111 pages. 1977

Vol. 5: David J. Clements, Brian D. O. Anderson
Singular Optimal Control: The Linear Quadratic Problem
V, 93 pages. 1978

Vol. 6: Optimization Techniques
Proceedings of the 8th IFIP Conference on Optimization Techniques, Würzburg, September 5–9, 1977
Part 1
Edited by J. Stoer
XIII, 528 pages. 1978

Vol. 7: Optimization Techniques
Proceedings of the 8th IFIP Conference on Optimization Techniques, Würzburg, September 5–9, 1977
Part 2
Edited by J. Stoer
XIII, 512 pages. 1978

Vol. 8: R. F. Curtain, A. J. Pritchard
Infinite Dimensional Linear Systems Theory
VII, 298 pages. 1978

Vol. 9: Y. M. El-Fattah, C. Foulard
Learning Systems:
Decision, Simulation, and Control
VII, 119 pages. 1978

Vol. 10: J. M. Maciejowski
The Modelling of Systems with Small Observation Sets
VII, 241 pages. 1978

Vol. 11: Y. Sewaragi, T. Soeda, S. Omatu
Modelling, Estimation, and Their Applications for Distributed Parameter Systems
VI, 269 pages. 1978

Vol. 12: I. Postlethwaite, A. G. J. McFarlane
A Complex Variable Approach to the Analysis of Linear Multivariable Feedback Systems
IV, 177 pages. 1979

Vol. 13: E. D. Sontag
Polynomial Response Maps
VIII, 168 pages. 1979

Vol. 14: International Symposium on Systems Optimization and Analysis
Rocquentcourt, December 11–13, 1978;
IRIA LABORIA
Edited by A. Bensoussan and J. Lions
VIII, 332 pages. 1979

Vol. 15: Semi-Infinite Programming
Proceedings of a Workshop, Bad Honnef, August 30 – September 1, 1978
V, 180 pages. 1979

Vol. 16: Stochastic Control Theory and Stochastic Differential Systems
Proceedings of a Workshop of the „Sonderforschungsbereich 72 der Deutschen Forschungsgemeinschaft an der Universität Bonn"
which took place in January 1979 at Bad Honnef
VIII, 615 pages. 1979

Vol. 17: O. I. Franksen, P. Falster, F. J. Evans
Qualitative Aspects of Large Scale Systems
Developing Design Rules Using APL
XII, 119 pages. 1979

Vol. 18: Modelling and Optimization of Complex Systems
Proceedings of the IFIP-TC 7 Working Conference
Novosibirsk, USSR, 3–9 July, 1978
Edited by G. I. Marchuk
VI, 293 pages. 1979

Vol. 19: Global and Large Scale System Models
Proceedings of the Center for Advanced Studies (CAS)
International Summer Seminar
Dubrovnik, Yugoslavia, August 21–26, 1978
Edited by B. Lazarević
VIII, 232 pages. 1979

Vol. 20: B. Egardt
Stability of Adaptive Controllers
V, 158 pages. 1979

Vol. 21: Martin B. Zarrop
Optimal Experiment Design for Dynamic System Identification
X, 197 pages. 1979

For further listing of published volumes please turn over to inside of back cover.

Lecture Notes in
Control and
Information Sciences
Edited by A.V. Balakrishnan and M. Thoma

50

M. Papageorgiou

Applications of Automatic Control Concepts to Traffic Flow Modeling and Control

Springer-Verlag
Berlin Heidelberg GmbH 1983

Author
Dr. Markos Papageorgiou
Dorsch Consult Ingenieurges. mbH
Postfach 210 243
D-8000 München 21

With 64 Figures

ISBN 978-3-540-12237-1 ISBN 978-3-540-39565-2 (eBook)
DOI 10.1007/978-3-540-39565-2

Originally published by Springer-Verlag Berlin Heidelberg New York Tokyo in 1983

2061/3020-543210

to Maria and to
my parents

Preface

Traffic flow on roads and freeways has become an important application field of automatic control theory in the last two decades. This monograph provides a review of the most recent research results achieved with respect to modeling, identification, surveillance and control of road traffic and freeway traffic systems. Applications of well-known optimization and automatic control methods like linear programming, nonlinear programming, identification techniques, Kalman filtering, maximum principle, decomposition methods, hierarchical optimization and multilayer control structures to traffic control problems are extensively discussed. Since some of these techniques are briefly outlined in the appendices, reading of the main part of the book seems to be possible also for readers without an extensive control theoretic background. The book is of interest both for traffic engineers wishing to get familiar with traffic control methods and for systems and control engineers interested in applications of modern control techniques to this important real-life application field.

The monograph is based on a twenty-hours-lecture I held as a visiting professor at the Dipartimento di Elettronica, Politecnico di Milano, Milan, Italy, in the spring 1982. The aim of the lecture was to provide a review of interesting applications of automatic control concepts to traffic control problems. The lecture was organized for students of the fourth and fifth academic years.
I would like to thank Professor G.Guardabassi and Professor A. Locatelli for the invitation and for interesting discussions during the course.

A great part of the text is devoted to research results achieved during my stay at the Lehrstuhl für Steuerungs- und Regelungstechnik, Technische Universität München, Munich, Fed. Rep. of Germany from 1976 to 1982. I wish to thank Professor G.Schmidt for many valuable suggestions and helpful discussions during this period, as well as the Deutsche Forschungsgemeinschaft (German Research Foundation) and the Bundesministerium für Verkehr (Federal Ministry of Traffic) for their financial support. Last but not least I wish to acknowledge the support of Dorsch Consult, Munich, in the preparation of the manuscript.

Markos Papageorgiou

Contents

1. Motivation and scope

During the last two decades there has been a considerable growth of interest and research work dedicated to the solution of various transportation problems. Among these problems, development of control systems for vehicular traffic flow on roads and freeways is certainly one of the most important. Until about fifteen years ago, the discipline of traffic systems analysis and planning was being dealt by civil engineers. However, in the progress of research it was more and more recognized that efficient utility of existing or planned streets can only be achieved, if traffic flow is viewed as a controllable process governed by specific deterministic or stochastic laws. Motivated by this fact both theoreticians and practitioners of Control Systems Engineering became more and more interested in many aspects of traffic systems.

These developments were forced by the exponential increase of the number of vehicles and recognition of the fact,that overload problems could not any more be solved by an according increasing of the street networks. The underlying idea, which is common to most control engineering applications is to try to "move with finesse instead of brute force" /1/. In fact, recent developments in control theory on the one side and in digital computers technology on the other side provide the necessary theoretical and practical tools for a satisfactory and cheap solution of most traffic flow problems /2-6/.

Application of control concepts implies the development and design of a closed-loop control system consisting of 1. measurement devices, 2. estimation algorithm and 3. control strategy, as is schematically shown for a freeway system in figure 1.1. More or less accurate mathematical models of traffic flow are needed for the development of estimation and control algorithms. For this reason our main concern along this monograph will lie on mathematical modelling of traffic flow on the one side and estimation and control algorithms on the other side. Review of the considerable amount of literature

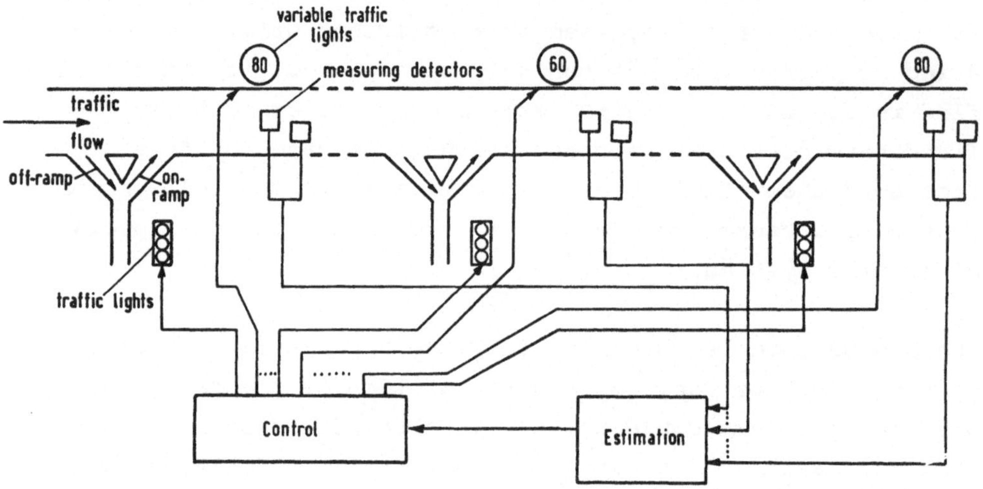

Figure 1.1 A freeway traffic control system.

concerned with the investigation of statistical properties of traffic flow /7/ will not be included, since it is of less importance for the development of control systems. In addition to general aspects applying to almost all urban traffic problems, the problem of controlling an oversaturated road network and the problem of controlling a constrained capacity freeway will be presented in some detail.

This monograph should not be understood as a comprehensive review of the existing literature on the subject, but as a tutorial introduction to the most important features of traffic flow process and of some control systems proposed by various researchers. Nevertheless, a rather long list of references is provided in order to make it possible for the interested reader to recover more details on specific topics. Excellent reviews of specific research areas can be found in /8-14/. The author himself was involved in the traffic control research during the last six years. As a more or less natural phenomenon, research work he participated covers a considerable amount of the present

The next section presents some quite general traffic flow modeling aspects as well as specific mathematical models for freeways and road intersections. In section 3 a specific identification and parameter estimation problem for a nonlinear freeway traffic model is outlined. State estimation of non-linear freeway traffic models by use of an extended Kalman Filtering is discussed in section 4. On the basis of the mathematical models presented in section 2, some steady-state and dynamic optimal control problems with respect to both freeway and road traffic are given in section 5. The important problem of controlling traffic flow on freeway corridors is briefly outlined, as well. Section 6 is concerned with a general presentation of the multilevel approach to the solution of dynamic nonlinear optimal control problems and its application to traffic control problems. The multilayer approach and its application to traffic control problems is described in section 7. Some concluding remarks as well as directions of future research are given in a final section 8.

2. Traffic flow models: a critical survey

The availability of adequate mathematical models describing traffic flow phenomena is an important prerequisite for application of modern control theory to traffic control problems. Estimation of traffic variables and development of efficient control strategies is not possible in absence of a certain degree of knowledge about the main "physical" laws governing the processes behaviour. The main purpose of this section is to describe the basic features of traffic flow and discuss several mathematical models which have been proposed in the past by various researchers.

2.1 General considerations

For the development of control strategies it is reasonable to make use of macroscopic mathematical models, i.e. models regarding traffic flow as a fluid with specific characteristics. On the other hand, it seems to be a natural way to begin with the investigation of the single elements constituting traffic flow. In that case, we have to consider m i c r o s c o p i c models describing the movement of a vehicle or a string of vehicles.

a) Microscopic models

Microscopic or car-following models consider each vehicle and its driver acting as a distance regulator on an individual basis. Car-following models can consider a whole string of vehicles following each other in a single lane. They take explicitly into account the reaction time (which should be understood as an average reaction time) of the drivers to outside stimuli. Such a stimulus for a driver could be a slowing down or a speeding up of the preceding vehicle, which would force him to hit the brake or the accelerator in order to keep up with the flow of vehicles.

The first microscopic models known are due to Reuschel /15/ and Pipes /16/. They describe the phenomena of the motion of pairs of vehicles following each other by the expression

$$x_n - x_{n+1} = L + S\,\dot{x}_{n+1} \qquad (2.1)$$

where x_{n+1} is the position of the vehicle n+1 considered and x_n is the position of the preceding vehicle as shown in figure 2.1. In this formula it is assumed that each driver maintains a separation distance proportional to the speed of his vehicle plus a distance L, which is the distance headway at standstill ($\dot{x}_{n+1} = \dot{x}_n = 0$) including the length of the lead vehicle. Differentiating eqn. (2.1) we obtain

$$\ddot{x}_{n+1} = \frac{1}{S}\,(\dot{x}_n - \dot{x}_{n+1}) \qquad (2.2)$$

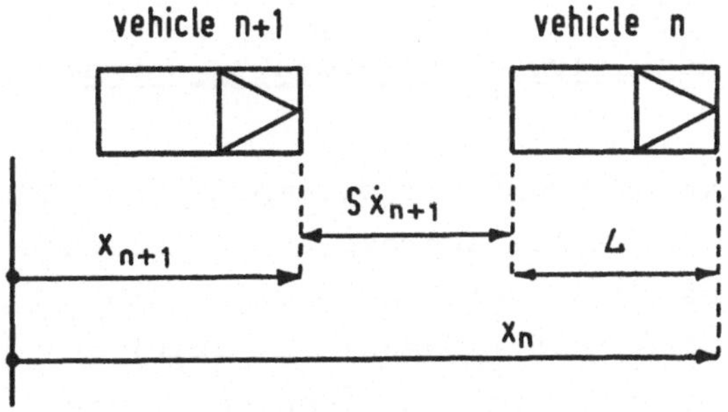

Figure 2.1 A pair of vehicles.

which means that acceleration or deceleration of a vehicle is proportional to its relative speed to the preceding vehicle. Defining a sensitivity factor $\lambda = 1/S$ and introducing a time lag τ of response to the stimulus /17/ we obtain from eqn.(2.2)

$$\ddot{x}_{n+1}(t + \tau) = \lambda \; [\dot{x}_n(t) - \dot{x}_{n+1}(t)] \tag{2.3}$$

which is generally referred to as the basic equation of car-following models. More accurate results can be achieved, if the sensitivity factor λ is expressed by use of the formula

$$\lambda = \lambda_o \; \frac{\dot{x}_{n+1}(t + \tau)^m}{[x_n(t) - x_{n+1}(t)]^l} \tag{2.4}$$

where λ_o is a constant and l, m are integer exponents. Various nonlinear car-following models are characterized by pairs of specific values (l,m). For instance, the linear model (2.3) is characterized by (0,0). Microscopic models are primarily used for simulation studies /18/.

b) The macroscopic, steady-state volume-density characteristic.

A macroscopic description of traffic flow implies the definition of adequate flow variables expressing the average behaviour of the vehicles at a specific location and time instant. We define the traffic density $\rho(x,t)$ as the number of vehicles per unit length [veh/km], the vehicles' mean speed $v(x,t)$ [km/h] and the traffic volume $q(x,t)$ as the number of vehicles passing a specific location in a time unit [veh/h] (1).

---

(1) Clearly, all these variables should be understood as mathematical abstractions, since they cannot have any physical meaning for infinitesimal dx, dt.

For a homogeneous traffic flow, it can be shown on the basis of microscopic considerations that the relation holds

$$q = \rho \cdot v \qquad (2.5)$$

which has its direct analogon in hydromechanics. Eqn.(2.5) approximately describes traffic flow also for inhomogeneous traffic conditions and is therefore included in most macroscopic models.

A real specific property of traffic flow, which does not have its analogon in any other fluid flow is the fact that traffic mean speed monotonically decreases with increasing density, as was found out by inspection of several sets of speed-density measurements.
As early as 1935, Greenshields /19/ hypothesized that a linear relationship existed between speed and density (figure 2.2)

$$v = v_f\left(1 - \frac{\rho}{\rho_{max}}\right) \qquad (2.6)$$

where v_f is the free speed and ρ_{max} is the jam density.

It is very interesting to state that the macroscopic relation (2.6) is a direct consequence of the linear microscopic model (2.3) for (l = 2, m = 0). Let us assume homogeneous and steady-state conditions ($\tau = 0$) and introduce the mean spacing s between vehicles

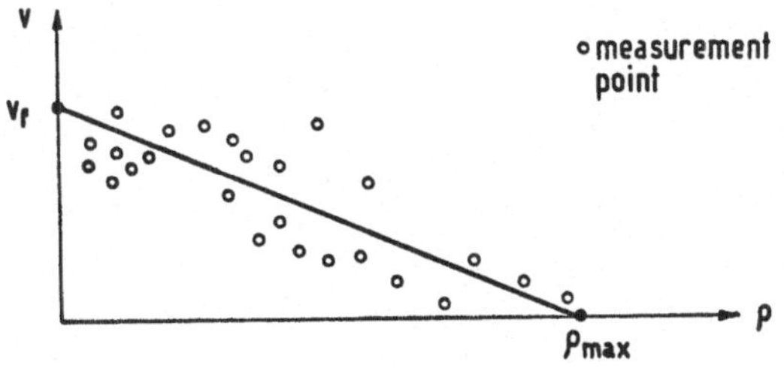

Figure 2.2 - The macroscopic, steady-state speed-density characteristic.

$$s \hat{=} x_n - x_{n+1} . \tag{2.7}$$

We recognize that

$$s = 1/\rho \quad \text{and} \quad L \hat{=} 1/\rho_{max} . \tag{2.8}$$

Integration of eqn. (2.3) then yields

$$v = -\frac{\lambda_o}{s} + b = -\lambda_o \rho + b \tag{2.9}$$

where b is an integration constant. If we assume that for $\rho = \rho_{max}$ the mean speed becomes zero, we get

$$b = \lambda_o \rho_{max} . \tag{2.10}$$

Setting $v_f = \lambda_o/\rho_{max}$, eqn. (2.9) becomes identical to equation (2.6). For different values of (l,m) we get different formulae for the steady-state speed-density characteristic (see /20/ for a detailled review). Clearly, once one decides to use macroscopic models, the best approach should be to determine the steady-state characteristic on the basis of macroscopic measurement sets. For example, one can prescribe the mathematical structure of the relationship and look for the parameter values leading to the best fitted curve in the minimum quadratic sense.

Once the speed-density characteristic has been determined, we can use eqn. (2.5) to determine the volume-density characteristic. For example, in the case of the linear formula (2.6) we get

$$q = \rho \cdot v = v_f \cdot \rho(1 - \frac{\rho}{\rho_{max}}) \tag{2.11}$$

which is shown in figure 2.3. We notice that traffic volume is increasing for increasing density until a maximum q_{max} is achieved for a critical density value ρ_{cr}. Further increase of traffic density leads to a decrease of traffic volume until $q(\rho_{max}) = 0$ is reached. This is a common phenomenon of all traffic systems and is due to safety considerations requiring a safe spacing adjustion between vehicles depending upon the vehicles speed.

A specific property of road traffic having to do with its dynamic behaviour results from the fact that distance regulation is carried out by car drivers. Extensive measurements

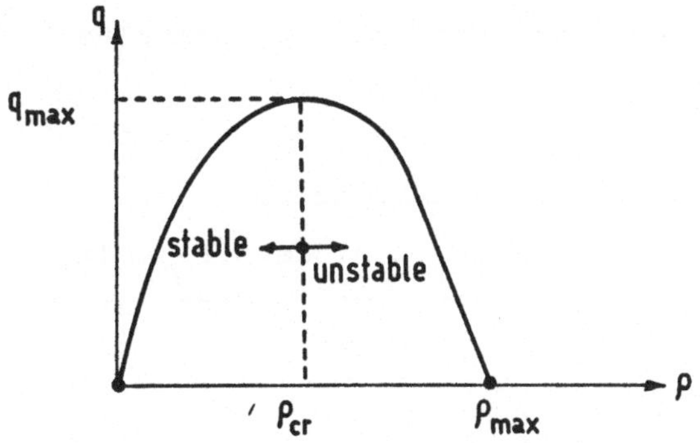

Figure 2.3 - The steady-state volume-density characteristic.

show that men acting as distance regulators lead to an unstable traffic flow when traffic density becomes overcritical ($\rho > \rho_{cr}$). Instability means here that, once an overcritical density has occured, the traffic flow becomes rapidly and without any obvious reason more and more congested, until values in the vicinity of the jam density are reached (stop-and-go traffic region). In terms of the microscopic models,instability refers to an increase in the amplitude of a perturbation as it propagates down a string of cars, finally leading to a car stoppage. Vehicle systems including a stable automatic distance regulator /21;22/, such as automated guide-way transit systems can operate at any point of the volume density characteristic without becoming unstable in the above sense.

2.2 Freeway traffic models

Let us now consider the special case of a long multilane freeway with many off-ramps and on-ramps. We are interested in developing a mathematical model of the traffic flow describing the dynamic evolution of traffic variables along the freeway.

a) Models based on the conservation equation

If we regard traffic flow as a fluid of density $\rho(x,t)$ and volume $q(x,t)$, then we may write the fundamental equation of conservation of matter

$$\frac{\partial \rho}{\partial t} + \frac{\partial q}{\partial x} = r - s \tag{2.12}$$

where $r - s$ is the on-ramp/off-ramp source term. In order to develop optimal control strategies, it is much more convenient to consider ordinary differential equations.
For this reason, we subdivide the freeway into a number N of sections with lengths δ_i, $i = 1,\ldots,N$, each having at most one on-ramp and one off-ramp. Furthermore, we introduce space-discretized traffic variables holding over a section i (figure 2.4)

- $\rho_i(t)$: number of vehicles in the freeway section no.i divided by the length δ_i of the section.
- $v_i(t)$: mean speed of vehicles in the freeway section no. i.

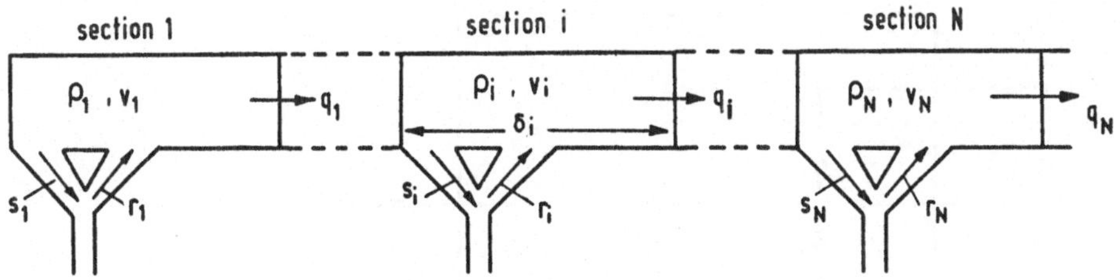

Figure 2.4 - A Freeway System subdivided into sections.

$q_i(t)$: number of vehicles leaving section i in a time unit.

$r_i(t)$, $s_i(t)$: on-ramp, off-ramp volumes of section i (if any).

With these variables, a space discretized form of eqn.(2.12) can be given :

$$\dot{\rho}_i(t) = \frac{1}{\delta_i} \left[q_{i-1}(t) - q_i(t) + r_i(t) - s_i(t) \right]. \qquad (2.13)$$

It is obvious, that in spite of the space discretization, eqn. (2.13) holds exactly for each section i.

If a steady-state model is needed, one obtains from eqn. (2.13) by setting $\dot{\rho}_i(t) \equiv 0$:

$$\bar{q}_i = \bar{q}_{i-1} + \bar{r}_i - \bar{s}_i \qquad (2.14)$$

where bar denotes steady-state values. A signal flow diagram for this simple model which we will call the model A is shown in figure 2.5

Let us now introduce some origin-destination-rates α_{ji} expressing the decimal fraction of vehicles entering the freeway at on-ramp j, which pass through section i. If the α_{ji}

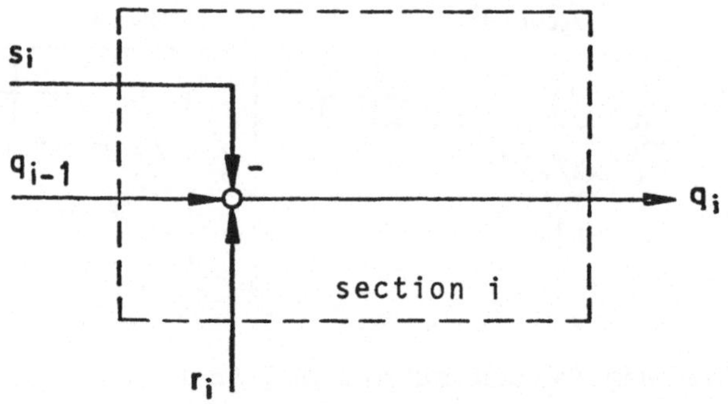

Figure 2.5 - Signal flow diagram for steady-state model A.

have constant values, then by definition

$$q_i = \sum_{j=1}^{i} \alpha_{ji} r_j, \quad 0 \leq \alpha_{jN} \leq \alpha_{jN-1} \leq \ldots \leq \alpha_{jj+1} \leq 1. \quad (2.15)$$

Substituting eqn. (2.15) in eqn. (2.14) we obtain

$$s_i = \gamma_i q_{i-1} \quad (2.16)$$

with the off-ramp volume rates γ_i given by

$$\gamma_i = \frac{\sum_{j=1}^{i} (\alpha_{j,i-1} - \alpha_{ji}) r_j}{\sum_{j=1}^{i-1} \alpha_{ji} r_j} . \quad (2.17)$$

Control strategies derived by use of this simple model will be discussed in section 5.

b) Extension by use of the volume-density characteristic

Conservation equation (2.12) resp. (2.13) can be extended to become a complete model of traffic dynamics by use of the nonlinear volume-density characteristic discussed in section 2.1b). We will consider both the original and the discretized form of the conservation equation.

By use of the partial differential equation (2.12) and a volume-density characteristic $q = q\ (\rho)$, Lighthill and Whitham /23/ and Richards /24/ derived some fundamental results which have been widely used for simulation /25/, surveillance /26/ and control /27/ of traffic flow. These results are based on the theory of kinematic waves /28/ and are best demonstrated in the volume-density characteristic shown in figure 2.6

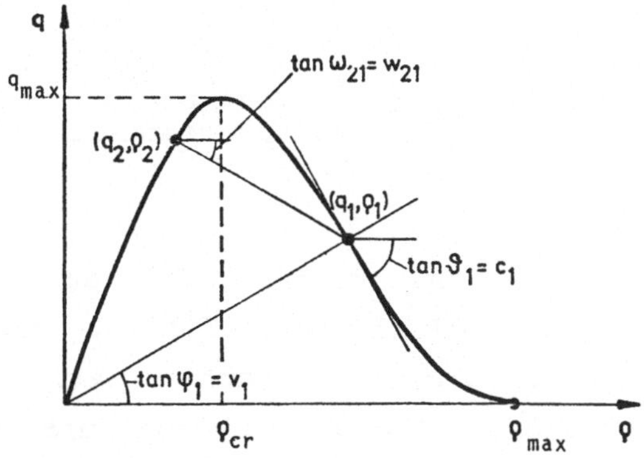

Figure 2.6 - Wave theory results.

In view of eqn. (2.5) we notice, that for a given traffic state point (q_1, ρ_1), the according mean speed v_1 can be interpreted in the volume-density characteristic, figure 2.6, as the slope of the radius vector from the origin leading to the point (q_1, ρ_1).

Suppressing the source term in eqn. (2.12) and assuming $q = q(\rho)$ we can write

$$\frac{\partial \rho}{\partial t} + c\,\frac{\partial \rho}{\partial x} = 0\ , \qquad c = \frac{\partial q}{\partial \rho}\ , \tag{2.18}$$

which has the solution

$$\rho(x,t) = F(x - ct) \tag{2.19}$$

where F is an arbitrary function. Eqn. (2.19) implies that inhomogeinities such as changes of traffic variables propagate along a stream of traffic at constant speed c. For example, a small change of density propagates with speed $\partial q/\partial \rho$, which can be interpreted as the slope of the tangent to the $q(\rho)$ - curve at a specific operating point (figure 2.6). Obviously, the following relations hold

$$c \leqslant v \quad \forall \rho \in [0, \rho_{max}] \tag{2.20}$$

$$c \geqslant 0 \quad \forall \rho \in [0, \rho_{cr}] \quad \text{(wave propagation in downstream direction)} \tag{2.21}$$

$$c < 0 \quad \forall \rho \in (\rho_{cr}, \rho_{max}] \quad \text{(wave propagation in upstream direction)} \tag{2.22}$$

These relations make clear that traffic state at a specific freeway location can be influenced either by the current traffic situation in upstream direction, which will occur when inequality (2.21) holds, or by the current traffic situation in downstream direction, namely when inequality (2.22) is valid. As a consequence, the signal flow direction in a mathematical traffic model describing traffic phenomena both for congested and uncongested traffic cannot be unique. Hence, the steady-state model of figure 2.5 described in section 2.2.a) is only valid on the left-hand side of the volume-density characteristic, because it includes only signal flows going in downstream direction.

Lighthill and Whitham also investigated the propagation of discontinuities, which continuous wave forms may develop due to overtaking of slower waves by faster ones (figure 2.7). We call the discontinuity a shock wave. The law of motion of shock waves is derived from conservation considerations. If the flow state is characterized by (q_2, ρ_2) on the upstream side and (q_1, ρ_1) on the downstream side and the shock wave is moving with speed w then the number

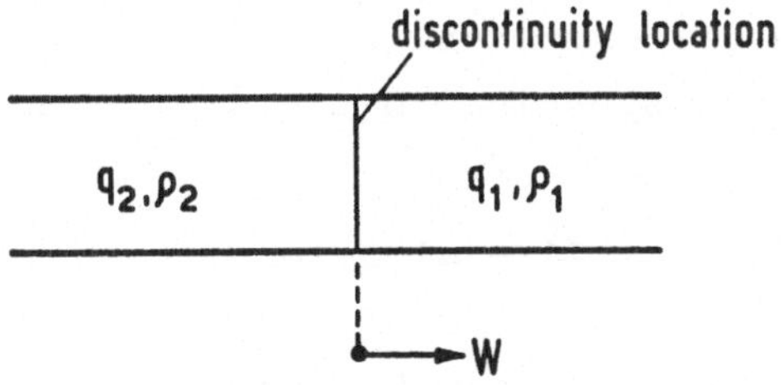

Figure 2.7 - Overtaking of slower vehicles (q_1, ρ_1) by faster ones (q_2, ρ_2)

of cars per time unit passing it are either $q_1 - w\rho_1$ or $q_2 - w\rho_2$. This gives the velocity of the shock wave as

$$w = \frac{q_1 - q_2}{\rho_1 - \rho_2} \tag{2.23}$$

which can be interpreted as the slope of the chord joining two points on the volume-density characteristic which represent conditions ahead of and behind the shock wave (figure 2.6).

Equations (2.19) and (2.23) represent the main results of the theory of kinematic traffic waves. The main simplification included limiting the accuracy of these results is the neglect of dynamic effects in the volume-density relationship. In other words, it has been assumed that any change of traffic density at a specific location is instantaneously followed by the corresponding change of traffic volume. Consequences of this assumption and possible improvements of the model accuracy are discussed in the section 2.2.d).

In order to enable the formulation of an optimal control problem on the basis of above assumptions, the volume-density characteristic is expressed in terms of the space discretized variables. For this purpose, the traffic volume between two freeway sections is expressed as a weighted sum of the traffic volumes corresponding to the densities of the sections

$$q_i(t) = \alpha . q\,[\rho_i(t)] + (1 - \alpha)\, q\,[\rho_{i+1}(t)] \quad (2.24)$$

where $0 \leqslant \alpha \leqslant 1$ is an appropriate weighting factor.

Equations (2.13) and (2.24) constitute a traffic flow model which we call model B. The corresponding flow diagram is given in figure 2.8. Figure 2.8 shows that model B includes signal flows both in upstream and in downstream direction and thus it fulfills a necessary condition for a reasonable description of both congested and uncongested traffic. However, unlike eqn. (2.13), eqn. (2.24) can only be considered as an approximation of real traffic phenomena.Its usefulness depends mainly upon the choice of the segment length /29/. Thus, too long or too short segment lengths may lead to a totally inadequate description of freeway traffic behaviour.

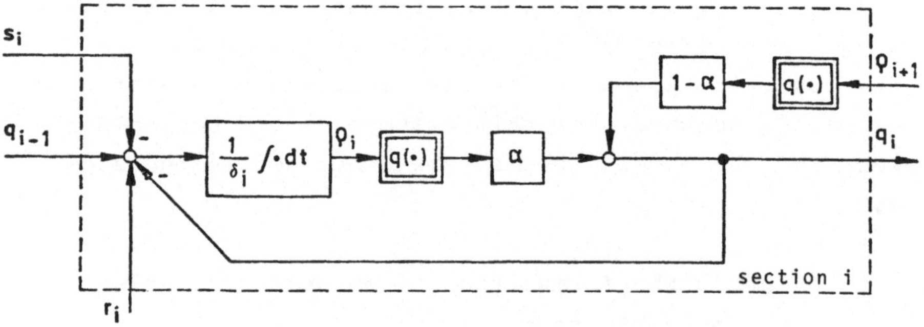

Figure 2.8 - Signal flow diagram of the nonlinear dynamic traffic flow model B.

c) A simplified dynamic approach

In view of the simple steady-state model given by eqn. (2.14) and its signal flow diagram, figure 2.5, it seems plausible to try to extend it just by adding a time delay τ_i as shown in figure 2.9. The new model obtained, which we will call model C, can be viewed either as a dynamic extension of the steady-state model (2.14) or as a simplified version of model B. In fact, it lies somewhere between the models A and B, since it has common features with both of them.

From figure 2.9, it is obvious that τ_i represents the mean travel time through the i-th freeway section. τ_i in general depends upon the mean speed in the same section. However, as long as traffic density remains undercritical , it can be assumed as being approximately constant, i.e. only depending upon the sections geometry. In view of the results of section 2.2.b), this corresponds to the simplifying assumption

$$v_i = c_i = \text{const.} \qquad \forall\, \rho \in [0, \rho_{cr}]$$

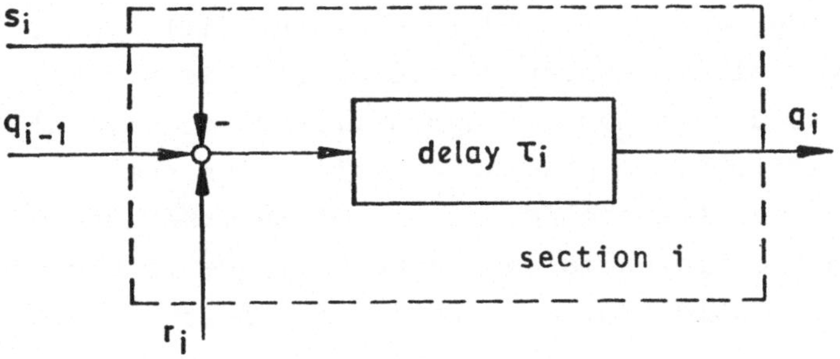

Figure 2.9 - Signal flow diagram of the dynamic, linear model C.

which means that the left-hand side of the volume-density characteristic is approximated by a straight line. A full analytical description of model C can be found in /30/. Let us summarize its main features:

(i) Model C is dynamic and still linear. As a consequence, improved control strategies compared to steady-state ones can be derived by its use, as will be further discussed in section 5. Since it is linear, well-known linear programming algorithms can be used.

(ii) As can be seen by inspection of figure 2.9, the signal flow direction of model C is unilateral (in downstream direction). That means, that no congested traffic situation can be described by the model and hence, if, due to an unexpected event, density becomes overcritical, it can not be used for the elimination of the occuring congestion.

d) Consideration of the mean speed dynamics

The freeway traffic model B shown in figure 2.8 has been found to be the most accurate mathematical model discussed up to this point. However, there are some significant traffic phenomena which can not be explained by this model:

(i) It has already been mentioned in section 2.1b) that a steady-state volume density characteristic is usually determined as a mean squares approximation of several measured traffic state points. At this point the questions arise: Are the deviations of measured state points from the approximating curve only due to stochastic effects? Could a dynamic relation provide a more accurate approach?

(ii) Instability of traffic flow on the right-hand side of the volume-density characteristic is not reproduced by model B, if a reasonable section length is chosen.

This is an obvious contradiction to known characteristic road traffic phenomena. Again the question arises: How should the traffic model be modified in order to include the mentioned instability effects with a reasonable degree of accuracy?

These two significant shortcomings of model B provide the justification for a dynamic extension leading to an improved description of traffic flow.

The improved model is due to Payne /31/ and has been developed on the basis of microscopic and empirical considerations. The basic idea consists in replacing the static relationship $q = q(\rho)$ of model B by a dynamic one, in which mean speed v is also involved. The improved model describes the dynamic evolution of density, volume and mean speed in freeway segments of 500-1000 m in length. In order to emphasize the fact that the space discretization intervalls for the improved model must be much shorter than for the previous models we will speak of freeway segments of lengths Δ_j, $j=1,\dots,n$. It is supposed that a freeway section consists of several segments.

Let us begin with eqn. (2.9) without neglecting the reaction time :

$$v(x,t + \tau) = -\lambda_o \rho(x,t) + b = v_e(\rho) \tag{2.25}$$

where v_e denotes the mean speed obtained for $\tau = 0$. since

$$v(x,t + \tau) = v(x, t) + \tau \frac{dv(x,t)}{dt} \tag{2.26}$$

we obtain

$$\dot{v} = -\frac{1}{\tau} [v - v_e(\rho)] = \frac{\partial v}{\partial t} + v \frac{\partial v}{\partial x} . \tag{2.27}$$

A slight modification of this procedure and subsequent space discretization finally leads to the following differential equation for segment mean speeds (see /31, 32/ for details)

$$\dot{v}_j = (v_{j-1} - v_j)\,\frac{v_j}{\Delta_j} + \frac{1}{\tau}(v_e(\rho_j) - v_j) - \frac{\rho_{j+1}-\rho_j}{\rho_j} \cdot \frac{\nu}{\tau \,.\, \Delta_j} \qquad (2.28)$$

where ν, τ are constant parameters. Though the procedure of obtaining macroscopic models seems to be straightforward, it should be noted that application of the resulting model to a multilane freeway situation represents a considerable extrapolation of the original area of application intended for the progenitor car following models. For instance, lane changing occuring in multilane traffic is not considered in microscopic models. For this reason, a validation procedure should follow after a reasonable structure of the model equations has been specified. Such a validation procedure will be described in section 3.

In any case the structure of eqn. (2.28) seems to be plausible from a macroscopic point of view. The first term at the right-hand side of eqn. (2.28) considers the influence of the incoming traffic on the mean speed evolution in segment j. The second term includes the speed-density characteristic as a "desired" value according to the current density ρ_j. The third term represents the influence of traffic density downstream on the mean speed evolution.

Traffic density is described by the known conservation equation (2.13). For the description of traffic volume between two adjacent segments a weighted mean form of eqn. (2.5) analogously to eqn. (2.24) is used

$$q_j = \alpha .\ \rho_j(t)\ .\ v_j(t) + (1-\alpha)\rho_{j+1}(t)\ v_{j+1}(t) \qquad (2.29)$$

Figure 2.10 shows the signal flow diagram of this accurate model, which we call model D. For an appropriate choice of its parameters,the model describes with satisfactory accurary instability phenomena for overcritical density

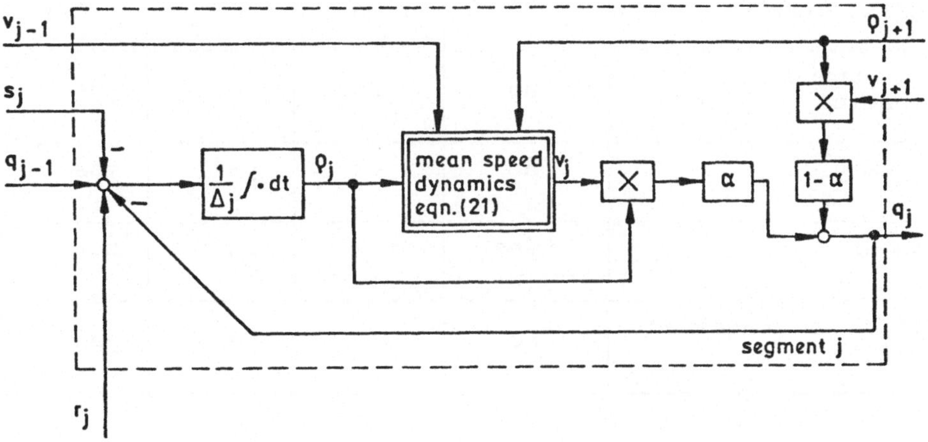

Figure 2.10 Signal flow diagram of traffic model D considering mean speed dynamics.

as we will see in section 3.

e) Summary of freeway models

In this section 2.2 we have discussed the four most important groups of macroscopic freeway traffic models with respect to their structural characteristics and modelling accuracy. The main statements of this unifying presentation are summarized in table 2.1.

Properties \ Models	A fig.2.5	C fig.2.9	B fig. 2.8	D fig.2.10
Conservation eqn.	yes	yes	yes	yes
Volume-density characteristic	no	simplified	yes	yes
Mean speed dynamics	no	no	no	yes
Dynamic	no	yes	yes	yes
Linear	yes	yes	no	no
Valid for congested traffic	no	no	yes	yes

Table 2.1 - Summary of the properties of the four freeway traffic models listed in an order of increasing accuracy.

2.3 Road traffic models

We will now proceed to the description of traffic phenomena at signalized intersections. It should be noticed, that the original reason why traffic lights have been introduced at city intersections have been safety considerations rather than control purposes. However, once traffic lights are installed, there are many ways of timing the green-red light periods and there is certainly a better way of doing it under specified conditions and a specified performance criterion. In other words, controlling an oversaturated intersection can be formulated as an optimization problem on the basis of mathematical models, which will be presented in this section.

Mathematical models of traffic at road intersections are quite different than the ones of traffic flow on a long street described up to this point. This is due to the fact, that attention is focused on the intersection phenomena, built up of queues etc. rather than on the process of traffic flow.

a) The oversaturated intersection

As long as vehicles arriving at an intersection during the red light period are served during the next green light period, no problem arises. However, particularly during rush hours, when the queues of vehicles at the intersection grow at a faster rate than the throughput, we have the case of an oversaturated intersection. A mathematical model of an over saturated intersection was developed by Gazis and Potts/33/ along with some preliminary analysis of the problem.

Two competing flows of vehicles, to be served by the intersection, are taken into account as shown in figure 2.11. The arrival rates of the vehicles in the two flows are d_1 and d_2. The maximum throughput rate, or "saturation flow", for both flows is s_1 and s_2 respectively. The effective green light phases in both directions are g_1 and g_2. The cycle time of the light interchange at the intersection will be denoted as t_c, and will be considered as being given. The average green light time required to serve all the cars arriving during a cycle t_c is

$$t_{gi} = t_c \frac{d_i}{s_i} \quad , \quad i = 1,2 \; . \tag{2.30}$$

When arrival flows d_1, d_2 increase so that

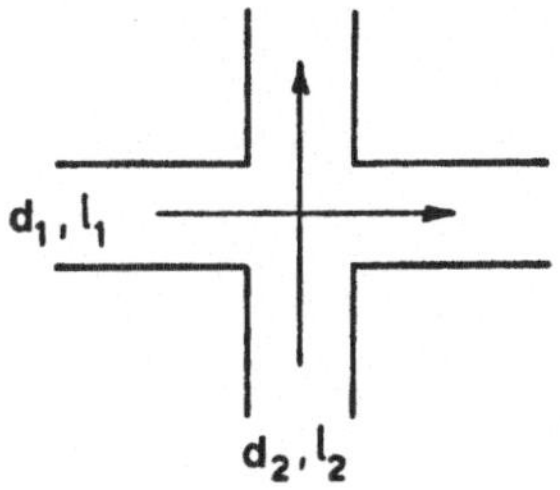

Figure 2.11 - A simple road intersection

$$t_{g1} + t_{g2} > t_o - L \tag{2.31}$$

or equivalently

$$\frac{d_1}{s_1} + \frac{d_2}{s_2} > 1 - \frac{L}{t_c} \tag{2.32}$$

we have the case of oversaturation. L is the total lost time for acceleration and clearing.

The queue lengths developing at the intersection are designated as the state variables of the system and denoted as l_1, l_2. The state equations are :

$$\dot{l}_i = d_i - s_i \frac{g_i}{t_c} \quad , \qquad i = 1,2 \ . \tag{2.33}$$

The first term on the right hand side of eqn. (2.33) represents the arriving flow, while the second term represents the flow served. The difference between the two terms is the rate of change of the queue length at the intersection. We have no influence on the rate of arrivals, however we can certainly specify the timing of the green light, which is included in the second term. Therefore we choose these terms as the control variables of the problem

$$r_i = s_i \frac{g_i}{t_c} \ , \qquad i = 1,2 \ . \tag{2.34}$$

But since

$$g_1 + g_2 = t_c - L \tag{2.35}$$

we obtain by setting eqn. (2.34) into (2.35)

$$\frac{r_1}{s_1} + \frac{r_2}{s_2} = 1 - \frac{L}{t_c} \tag{2.36}$$

or equivalently

$$r_2 = s_2(1 - \frac{L}{t_c}) + \frac{s_2}{s_1} r_1 \tag{2.37}$$

which means that we need only one control variable $r = r_1$. Setting eqns. (2.34) and (2.37) into (2.33) we obtain the following system of linear state equations :

$$\dot{i}_1 = d_1 - r \tag{2.38}$$

$$\dot{i}_2 = d_2 - s_2(1 - \frac{L}{t_c}) + \frac{s_2}{s_1} r \; . \tag{2.39}$$

These equations will provide a basis for deriving optimal control strategies for the signalized intersection (section 5).

In real networks, intersections are rarely as simple as the one shown in figure 2.11. Once turning movements are introduced, they become much more complex. Consider the one way intersection with turns as shown in figure 2.12.
Now, in traffic engineering practice, it is accepted that the proportions of turns to non-turning transits are fairly stable and it is possible to assign fixed values to these for different times of day. Thus a turning movement could

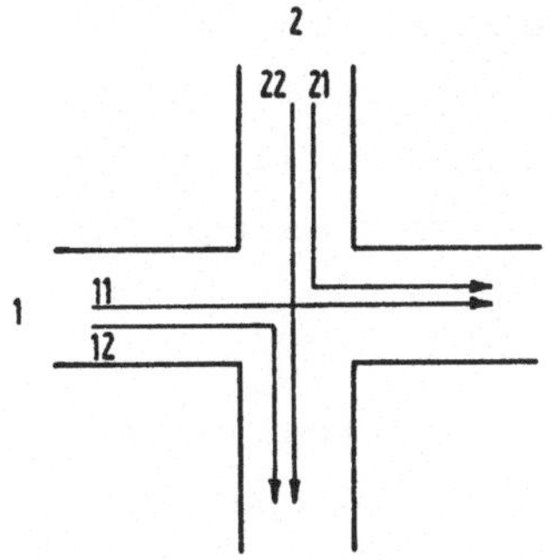

Figure 2.12: A complex over saturated intersection.

be thought as a discharge from a queue on the input arm. The queue may be real of fictitious. Real queues for turning movements exist when there is a separate turning lane. However it is still possible to consider a queue for turning with its associated saturation flow rate where there is no separate lane for turning, since for oversaturated networks average outflow from turning is usually constant.

The state equations of the intersection shown in figure 2.12 can be derived in analogy to those of a simple intersection. We have to distinguish between four different streams of vehicles. The control variable is set to

$$r = s_{11} \frac{g_1}{t_c} \tag{2.40}$$

and the state equations are given by

$$\dot{i}_{11} = d_{11} - r \tag{2.41}$$

$$\dot{i}_{12} = d_{12} - \frac{s_{12}}{s_{11}} r \tag{2.42}$$

$$\dot{i}_{21} = d_{21} - s_{21} \left(1 - \frac{L}{t_c}\right) + \frac{s_{21}}{s_{11}} r \tag{2.43}$$

$$\dot{i}_{22} = d_{22} - s_{22} \left(1 - \frac{L}{t_c}\right) + \frac{s_{22}}{s_{11}} r \quad . \tag{2.44}$$

In a similar way, mathematical models of several subsequent intersections can be developed. In that case, the arrival rates of the downstream intersections are directly related to the serving rates of the up stream intersections so that a coupled system of differential equations results /34/, see also section 6.2b).

For simulation studies, many mathematical models of traffic on an urban network have been developed, see /35/ for a review.

2.3b) Urban traffic networks as discrete-time point processes.

Recently, research work has been devoted to modelling, estimation and control of urban traffic networks on the basis of the theory of discrete-time point processes /36-38/. Without going into details, we will briefly outline the modeling procedure for the case of a single intersection. The idea behind this approach has been to enable the design of closed-loop control systems for traffic networks with moderate flow conditions (not oversatured).

Consider again the simple intersection shown in figure 2.13. Assume that there is one detector located N vehicle length from the stop line. The observed signal from the detector will be denoted by $n^a(t)$:

$$n^a(t) = \begin{cases} 1 & \text{if a vehicle is passing over the detector} \\ 0 & \text{otherwise.} \end{cases} \tag{2.45}$$

For simplicity, it is assumed that each vehicle produces exactly one pulse. Define

$$\lambda(k,t) = P_r \left[n^a(t) = 1 \text{ given that there are k vehicles in the queue and the time is t} \right] \tag{2.46}$$

$$k = 0,1,2,\ldots,N.$$

Assume that

$$\lambda(k,t) = \lambda(t) = \lambda_r U_u(t-\tau) + \lambda_g(1-U_u(t-\tau)) \tag{2.47}$$

$$\text{for } k = 0,1,2,\ldots,N-1$$

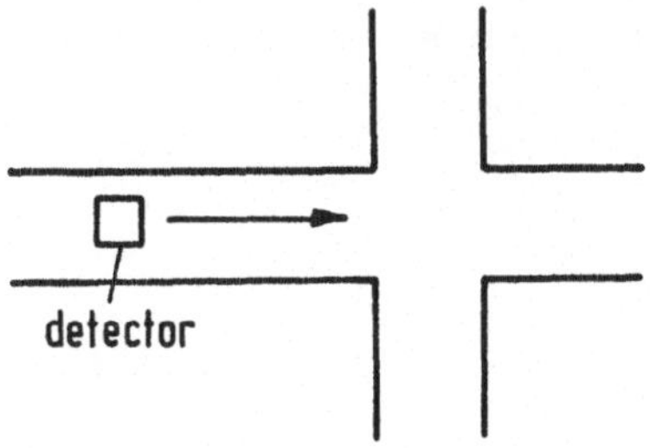

Figure 2.13 - A simple intersection with detector measurements.

and

$$\lambda(N,t) = 0 \tag{2.48}$$

where

$$U_u(t) = \begin{cases} 1 & \text{if upstream traffic signal is red} \\ 0 & \text{if upstream traffic signal is green} \end{cases}$$

and τ is the known average time for a vehicle to get from the upstream stop line to the detector.

It is important to understand the implications of the above assumptions. First, when the queue length contains N vehicles, no more vehicles can cross the detector, which leads to eqn. (2.48). Of course, in very heavy traffic flow conditions the queue may very well exceed N by a substantial amount. This limits the validity of the model to less than very heavy flow conditions.

Secondly, it has been assumed that the time required for the lead vehicle in the queue to proceed to the next detector is very predictable (always τ , see section 2.2.c)). This is again a reasonable assumption only in moderate traffic flow. Finally, the two level approximation in eqn.(2.47) is a quite usual assumption.

To complete the model, imagine there is a detector at the stop line producing an unobserved departure point process $n^d(t)$

$$n^d(t) = \begin{cases} 1 & \text{if a vehicle departs at time t} \\ 0 & \text{otherwise.} \end{cases}$$

Let

$$\mu(k,t) = P_r \; [\text{1 vehicle departs given k vehicles in queue at time t}].$$

Then

$$\mu(k,t) = \begin{cases} \mu \cdot (1-u_i(t)) & \text{for } k > 0 \\ 0 & \text{for } k = 0 \end{cases} \qquad (2.49)$$

where

$$u_i = \begin{cases} 1 & \text{if traffic signal at the intersection is red} \\ 0 & \text{if traffic signal at the intersection is green.} \end{cases}$$

This actually completes the model for queueing at one arm of the intersection. To see this, note that if the switch times of traffic signals are known

$$P_r\,[l(t+1) = j \;|\, l(t) = i\,] =$$

$$= \begin{cases} 0 & \text{for } j < i-1, \quad j > i+1 \\ \lambda(i,t)\,\mu(i,t) + (1-\lambda(i,t))(1-\mu(i,t)) & \text{for } i=j \\ \lambda(i,t)\;(1-\mu(i,t)) & j = i+1 \\ \mu(i,t)\;(1-\lambda(i,t)) & j = i-1 \end{cases} \qquad (2.50)$$

and

$$P_r\,[n^a(t) = 1 \,|\, l(t) = i] = \begin{cases} \lambda(t) & \text{for } i < N \\ 0 & \text{for } i = N \end{cases} \qquad (2.51)$$

where l is the queue length in the one arm of the intersection. Equations (2.50), (2.51) provide a basis for estimation and control of the intersection. It is straightforward to extend the model to complicated intersections.

3. Traffic flow models identification: a case study

Macroscopic traffic flow models derived in the previous section contain a number of parameters, the values of which should be specified before they can be used for the development of estimation and control algorithms.

These parameters are :

For the freeway traffic models

model A : α_{ij}

model B : α and parameters of the speed-density characteristic

model C : τ_i, α_{ij}

model D : α, τ, ν, and parameters of the speed-density characteristic.

For the road traffic models

oversaturated intersection: s_1, s_2, L

discrete-time point model : λ_g, λ_r, μ.

In this section, an off-line parameter estimation procedure for the freeway traffic model D is described |29, 39|.
On the basis of available measured traffic data, the identification procedure is formulated as a nonlinear parameter optimization problem which is solved my means of a nonlinear programming algorithm. The data sets used cover a large variety of different traffic situations, in order to ensure reliable and representative results.

3.1 Mathematical model equations

As already mentioned, the freeway traffic model D is considered in the identification procedure. Some modifications of the original form given in section 2.2.d) seemed to be reasonable.

The discrete-time version of the continuous time model of section 2.2.d) with time interval T, has been considered. The discrete-time variables $\rho_j(k)$, $v_j(k)$, $q_j(k)$ are defined at time instants k = 0,1,2,...,K. We consider a single freeway section consisting of n freeway segments without any on-ramps or off-ramps. The discrete-time model equations for a freeway segment are directly derived from eqns. (2.13), (2.28), (2.29) by use of a simple Euler-formula :

$$\rho_j(k+1) = \rho_j(k) + \frac{T}{\Delta_j} [q_{j-1}(k) - q_j(k)] \tag{3.1}$$

$$v_j(k+1) = v_j(k) + \frac{T}{\tau} [v_e(\rho_j(k)) - v_j(k)] + \frac{T}{\Delta_j} v_j(k) [v_{j-1}(k) - v_j(k)] - \frac{v}{\tau} \frac{T}{\Delta_j} \frac{\rho_{j+1}(k) - \rho_j(k)}{\rho_j(k) + \chi} \tag{3.2}$$

$$q_j(k) = \alpha . \rho_j(k) v_j(k) + (1-\alpha)\rho_{j+1}(k) v_{j+1}(k) \tag{3.3}$$

where a farther parameter χ was added in the last term of the right hand side of eqn. (3.2) in order to increase the accuracy of the model for low traffic densities. A fairly complex and accurate speed- density characteristic is used

$$v_e(\rho) = v_f [1 - (\frac{\rho}{\rho_{max}})^l]^m \tag{3.4}$$

with unknown parameters v_f, ρ_{max}, m, l.

Measurement sets provided by traffic detectors, which have been located at the both ends of the section and between the Jth and (J+1)st segment (figure 3.1) are available. The measurements consist of discrete-time values for traffic volume q and time mean speed w, which is the harmonic mean /40/ of the velocities of the individual vehicles passing over the detector. w is related to v through

$$w_j(k) = \alpha . v_j(k) + (1 - \alpha)v_{j+1}(k). \tag{3.5}$$

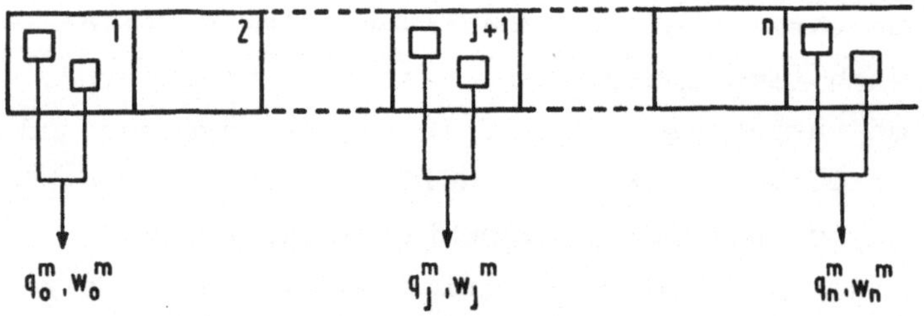

Figure 3.1 - Measurement locations on the freeway section.

In order to provide the values of $v_0(k)$ in eqn. (3.2) for the case j=1, we simply set

$$v_0(k) = w_0(k). \tag{3.6}$$

The values of $\rho_{n+1}(k)$ required in eqn. (3.2) for j=n are given by

$$\rho_{n+1}(k) = q_n(k)/ w_n(k). \tag{3.7}$$

Thus, the measured variables obtained by the detectors at the both ends of the freeway section are used as input variables, whilst q_J, w_J are considered as the system output. A justification of this choice can be easily given in view of the results of section 2.2b). We recall that in the case of low density, traffic flow in the considered section is mainly influenced by traffic conditions at the section's entry, whilst for high densities the influence of traffic con ditions at the section exit should be considered. Since the model is supposed to describe traffic dynamics in the whole density range, $0 \leqslant \rho \leqslant \rho_{max}$, traffic variables at both the entry and the exit of the section should be treated as input variables. On the other hand, the traffic variables q_J, w_J are completely reproducible by use of the model eqns. (3.3.) and (3.5) and can be considered as output variables. Consequently, a configuration with three sites of data collection shown in figure 3.1 is a suitable one for the proposed estimation prodedure.

In order to formalize the treatment, we introduce a state vector

$$\underline{x}^T = [\rho_1 \; v_1 \; \ldots \; \rho_n \; v_n] \quad , \; \underline{x} \in R^{2n}$$

an input vector

$$\underline{u}^T = [q_o \; w_o \; q_n \; w_n] \quad , \; \underline{u} \in R^4$$

an output vector

$$\underline{y}^T = [q_J \; w_J], \quad \underline{y} \in R^2$$

and a parameter vector

$$\underline{\beta}^T = [v_f \; \rho_{max} \; l \; m \; \alpha \; x \; \nu \; \tau] \quad , \; \underline{\beta} \in B \subset R^8$$

where B is a reasonably chosen subset of the parameter space. Furthermore, let $\underline{u}^m(k)$, $\underline{y}^m(k)$, $k = 1,\ldots,K$, be the time sequences of measured data collected from a real traffic flow, which contains transitions through the whole spectrum of possible density values in a representative manner.

Substituting eqn. (3.3) for $j = 1,2,\ldots,n-1$ and eqns. (3.4), (3.6) and (3.7) into eqns. (3.1) and (3.2) for the appropriate segment numbers, we obtain the following nonlinear dynamic state vector equation

$$\underline{x}(k+1) = \underline{f}\,[\underline{x}(k), \underline{u}(k), \underline{\beta}] \; . \tag{3.8}$$

Since the identification procedure is performed off-line, a reasonable initial condition can be given by inspection of the measured data

$$\underline{x}(0) = \underline{x}_o \; . \tag{3.9}$$

As an alternative approach, initial values of the state variables could be considered as additional unknown parameters. The output vector equation is built up by equations (3.3) and (3.5) taken for j = J and has the general form

$$\underline{y}(k) = \underline{g}\,[\underline{x}\,(k),\ \underline{\beta}] \tag{3.10}$$

3.2 The identification procedure

The estimation of the unknown parameters for the described system is a nontrivial task, since system equations are highly nonlinear in both the parameters and the state variables. The most common approach for the identification of nonlinear systems is the least squares output error method which minimizes the discrepancy between the model and the real process with respect to some quadratic output error functional. This approach has the additional advantage that it does not need any further a priori information about the probabilistic properties of the parameter values /41/ which are not available in our case.

In that sense, the parameter estimation problem may now be formulated as the following least squares output error problem : Given the time sequences of measured data

$$\underline{u}^m(k),\ \underline{y}^m(k),\ k = 1,2,\ldots,K$$

and the initial state $\underline{x}(0)$, find the set of parameters $\beta \varepsilon B$ minimizing the cost functional

$$J(\underline{\beta}) = \sum_{k=1}^{K} \| y(k) - y^m(k) \|_Q^2 \tag{3.11}$$

subject to eqns. (3.8) for $\underline{u}(k) = \underline{u}^m(k)$, and (3.10).

Q is a positive definite 2x2 matrix which was chosen to be

$Q = \text{diag}\ (\gamma, 1)$ with a weighting factor γ appropriately selected, for example

$$\gamma = \sigma_w^2 / \sigma_q^2 \qquad (3.12)$$

where σ_q^2, σ_w^2 are the variances of the stochastic components in the measured variables which cannot be modelled by the deterministic equations of the model. Here

$$\gamma = 0.001\ \text{km}^2 / \text{veh}^2$$

has been chosen.

A well-known approach to the solution of the formulated optimization problem is performed through formal extension of the state space by use of the equation

$$\underline{\beta}(k+1) = \underline{\beta}(k) \qquad (3.13)$$

and utilization of quasilinearization techniques /42/ for the solution of the resulting Two-Point-Boundary-Value-Problem (TPBVP). However, the equations of the TPBVP provide necessary conditions for a local minimum, whereas we are interested at the global minimum. Some preliminary investigations of the nonlinear optimization problem showed that there are in fact a lot of local minima distributed over the parameter space. Since no information about the possible location of the global minimum or about the number of existing local minima has been available, application of quasilinearization techniques should probably not lead to the global minimum. For exactly the same reasons, direct optimization methods, like gradient methods or steepest decent methods with numerical calculation of the gradient $\frac{\partial J^T}{\partial \underline{\beta}}$ have been excluded, as well.

In that sense, application of the Complex algorithm of Box /43/ seemed to be a reasonable approach. This algorithm does not require the calculation of derivatives of the cost functional. The algorithm starts with an initial set of points $\underline{\beta}^i$, i=1,2,..,12, which are randomly scattered throughout the admissible region in the parameter space. After each iteration step, the parameter set with the worst value of the cost functional is replaced by a new parameter set chosen according to an appropriate search routine. Of course, for each choice of a new parameter set, the value of the cost functional must be computed by a simulation run of the model equations driven by the measured inputs according to figure 3.2.

The Complex algorithm is more probable to find the global optimum or at least a "good" parameter set, since it starts with randomly scattered initial points and proceeds without using gradient calculations. The procedure is terminated when the points $\underline{\beta}^i$

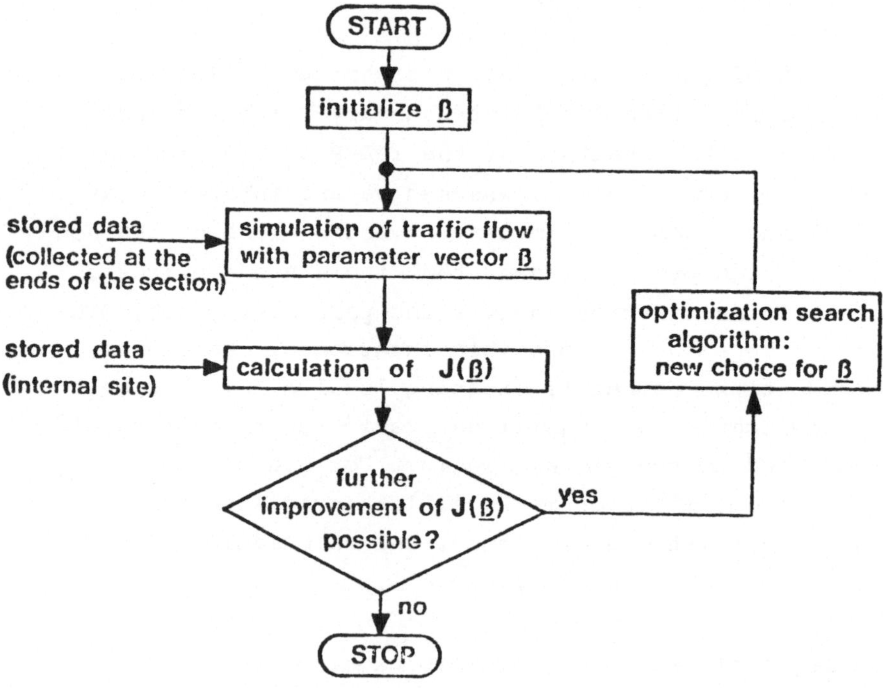

Figure 3.2 - Optimization structure

reach a sufficiently small region around the optimum so that no further improvement of the performance functional can be achieved by further iterations. Even with this algorithm, however, it is not easy to decide whether the global optimum has been actually reached. For this reason, it is useful to repeat the procedure with different sets of starting points.

3.3 Results

a) The optimal parameter set

For the execution of the outlined identification procedure a number of different data sets were available which were collected from a two-lane section of the Autobahn from Frankfurt to Basel. The section has a total length of 2650m which was subdivided into n = 5 segments of 500 m (j=1,2) and 550 m(j=3,4,5) length. Sensors for volumes and velocities are installed at both ends and within the section at a distance of 1000 m from the section's entry, i.e. behind the second segment, J=2. The available data sets contained a number of different traffic situations including free as well as congested traffic flow. From these, a representative set of 3 1/2 hours observation period was selected. In the last hour of this measurement set traffic became more and more crowded and finally collapsed.

By the identification procedure described in the preceding section, the set of optimal parameter values shown in table 3.1 was obtained. The optimization with the Complex algorithm of Box was performed on a Cyber 175 digital computer. Convergence was achieved after 148 iterations which took about 5 min computation time. The sampling time T was chosen to be 10 s, so that each simulation run over the 3 1/2 h real time period consisted of K = 1260 sampling time intervals. The optimization was carried out several times with different sets of starting points. In the

v_f	ρ_{max}	l	m	α	χ	ν	τ
123 km/h	200 veh/km	4,0	1,4	0,8	20 veh/km	21,6 km^2/h	0,01 h

Table 3.1 - The optimal parameter set.

most cases the parameter set of Table 3.1 resulted. As was mentioned above, this makes it rather likely that the absolute optimum was achieved.

In figures 3.3 and 3.4 the time responses of local mean speed $w_2(k)$ and volume $q_2(k)$ as generated by the calibrated model are presented together with the measured sequences $w_2^m(k)$ and $q_2^m(k)$ of the real traffic process for the last of the 3 1/2 hours. By inspection of the detailled results it can be seen that the model with the obtained parameter values reproduces with satisfactory accuracy the instability phenomena occuring at overcritical density values.

b) Transferability of the results

It is an interesting investigation to find out to which extend the performance of the identified model is sensitive with respect to the chosen measurement set. In order to answer this question, the model with the optimal parameters of table 3.1 was applied to a number of different data sets collected from the same freeway section. First, the performance functional (3.11) was evaluated for the parameter values of table 3.1. Then, the identification procedure described in section 3.2 was carried out for each new data set in order to find out which degree of improvement of the model's quality could be obtained by an individual parameter optimization.

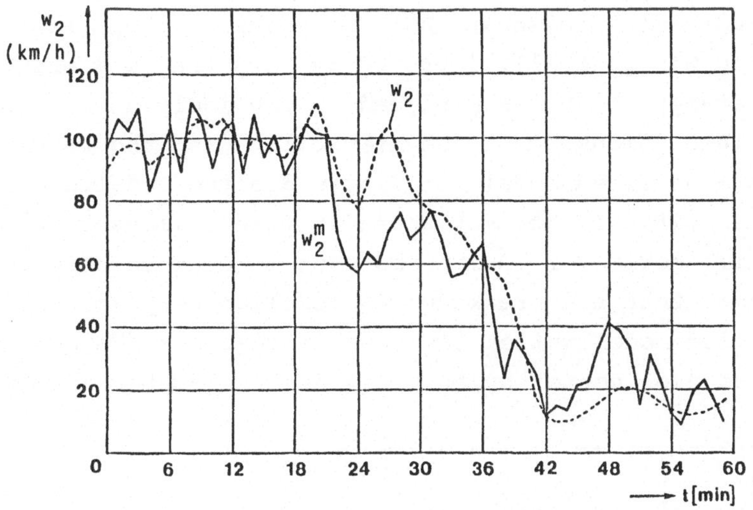

Figure 3.3 - Local mean speed $w_2(k)$: measurements (——) and model (---)

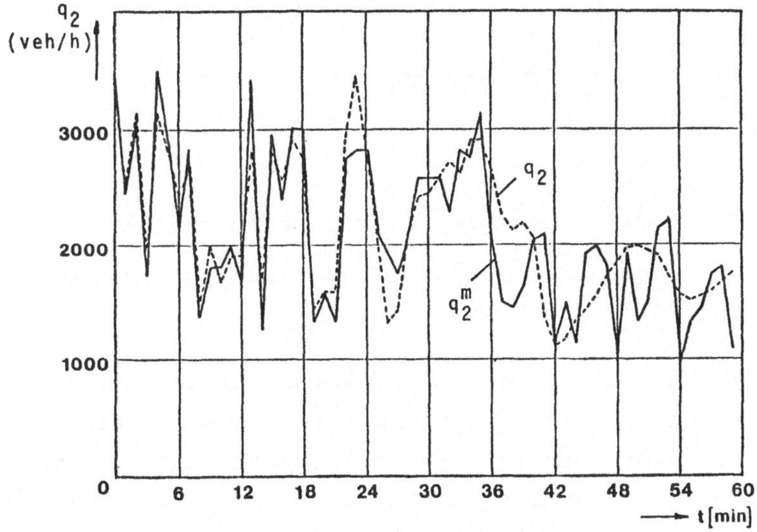

figure 3.4 - Volume $q_2(k)$: measurements (——) and model (---)

These investigations have shown that the value of the performance criterion was generally improved by less than 20% when the nominal parameter set was replaced by the individually optimized set. This demonstrates the flexibility and the transferability of the identified model. In figure 3.5 the results are depicted for a critical data set where traffic collapses only for a short time period of 10 min and then returns to normal flow. The curves show that the breakdown is modelled very accurately by the identified model whilst the special shape of the phenomenon is fitted even better by an individual parameter adaptation.

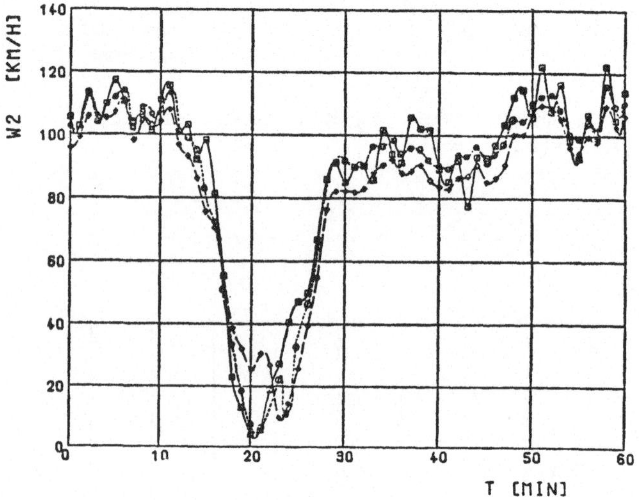

Figure 3.5 - Local mean speed $w_2(k)$: measurements (——) model with parameters of table 3.1 (---) and model with individually adapted parameters (....)

It is believed, that the specified parameter set is rather sensitive with respect to significant modifications of the section's geometry. This means that the results are not applicable to sections with unusual strong curves, uphill segments etc. Individual parameter estimation should be performed in such cases. Results have been also obtained concerning

- sensitivity with respect to parameter changes
- influence of sampling time and segment length
- possible simplifications of the model structure.

The full results are reported in /44/. As a conclusion, it could be stated that the model structure D presented in section 2.2.d) with appropriately selected parameter values can describe with a satisfactory accuracy dynamic traffic phenomena on freeways. A simulation package developed on the basis of this model is described in /45/.

4. Estimation of traffic flow variables on freeways

4.1 General considerations

As already mentioned with respect to the closed-loop control scheme of figure 1.1, the development of estimation algorithms reconstructing the current traffic state from specific measurements is an important part of an automatic control system. In this chapter, we will be concerned with the estimation of traffic variables on long roads (freeways) which has found considerable attention in the control literature.

Before we proceed to the development of estimation algorithms, it is important at this point to clarify the difference between the estimation of traffic variables and the identification procedure described in section 3. In view of the excellent results of the last section one may wonder about the necessity of a special estimation procedure. In fact, traffic state between two detector locations could be reproduced just by fitting the model equations with the measured values q_0, w_0, q_n, w_n as was done in section 3 (figure 4.1). It should be recalled, however, that this procedure implies

- known initial conditions
- accurate measurements.

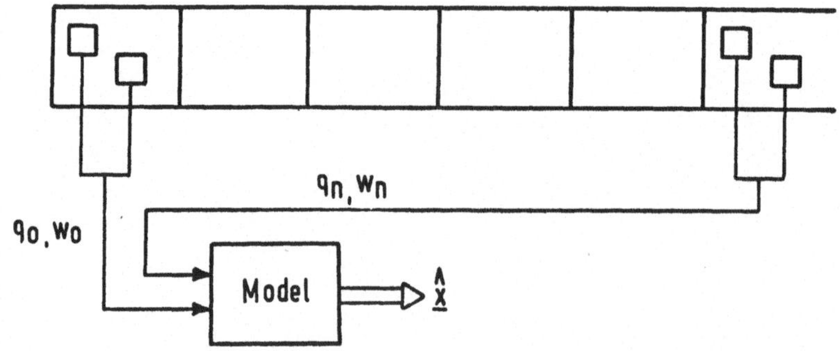

Figure 4.1 - State reconstruction with four input variables.

In an off-line procedure like the one of section 3, the a priori available measurment set can be cleared from detector errors and besides, an initial condition can be specified. In an on-line estimation, however, above implications are not fullfilled. Initial conditions are unknown and detector measurements are more or less inaccurate. For this reason, proceeding without some kind of comparison between process data and model results could lead to a totally inaccurate state reconstruction as will be shown in section 4.3. Well-known techniques like observers and filters use measured output variables in order to compare simulated and real data. In our case, the four measured variables could be interpreted both as input and as output variables as was discussed in section 3. Hence, we can use a part of them as input variables, in order to fit the model equations,and the rest as output variables for comparison purposes. As an example, the structure shown in figure 4.2 may result. The most appropriate selection of input resp. output variables from the given measurement set depends upon the method used as will be shown in this section.

Because of the stochastic effects involved in traffic measurements and traffic dynamics, Kalman filtering techniques have

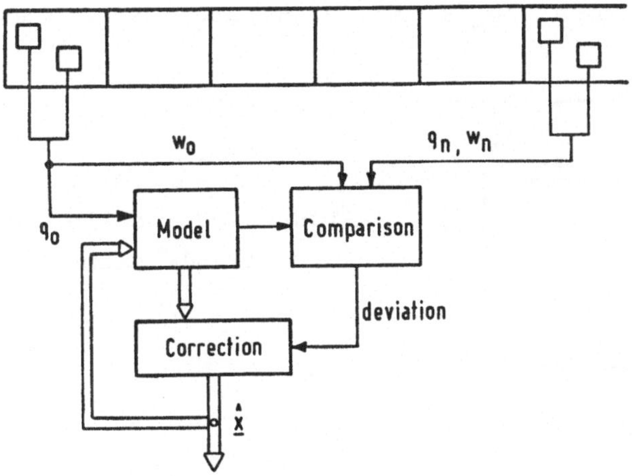

Figure 4.2 - State reconstruction with one input and three output variables.

been widely used by various researchers /44; 46-51/.
In the next two sections we will describe two alternative ways of estimating traffic variables. The problem formulation and solution equations of Kalman Filters are reviewed in Appendix 1.

4.2 State estimation for a single segment

a) Model equations

Most research work on estimation of freeway traffic variables considers a single segment with unique density and mean speed values. Detectors are supposed to be installed at both ends of the segment providing the observations

$$u_1(k) = q_o(k) + \zeta_1(k) \tag{4.1}$$

$$y_1(k) = q_1(k) + \zeta_2(k) \tag{4.2}$$

$$u_2(k) = w_o(k) + \zeta_3(k) \tag{4.3}$$

$$y_2(k) = w_1(k) + \zeta_4(k) \tag{4.4}$$

where $\zeta_i(k)$, $i = 1,...,4$ is the observation noise which is due to several measurement effects. The statistics of ζ_i are assumed to be known. $w_o(k)$ and $w_1(k)$ are the mean speeds

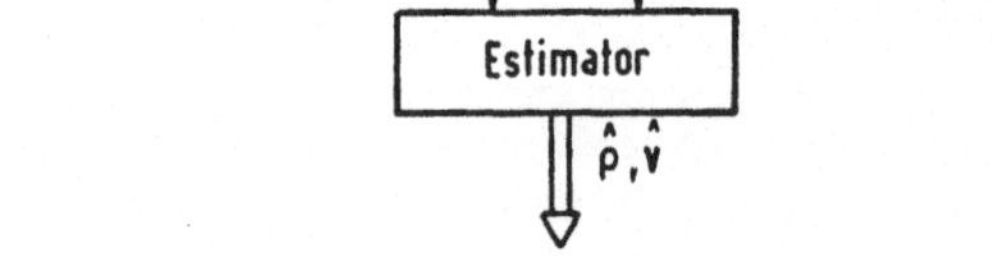

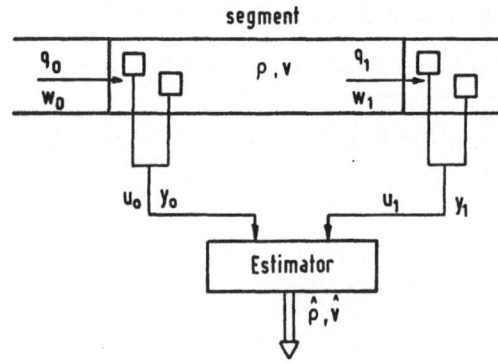

Figure 4.3 - State Estimation for a single segment.

of the vehicles entering resp. leaving the segment during a time interval. Following the development of /48, 49/ we will derive an estimation algorithm for density ρ and mean speed v in the segment, as shown in figure 4.3.

As is indicated by eqns. (4.1) - (4.4) we intend to use the measurements of q_o and w_o as input variables and the ones of q_1 and w_1 as output variables. Since a single segment is considered we obtain from eqns. (2.5) and (4.2)

$$y_1(k) = \rho(k) \cdot v(k) + \phi(k) + \zeta_2(k) \qquad (4.5)$$

where $\phi(k)$ is an additional zero mean noise term considering inaccuracies of eqn. (2.5). From equations (3.1), (4.1), (4.2) and (4.5) we obtain for the traffic density

$$\rho(k+1) = \rho(k) + \frac{T}{\Delta}\left[u_1(k) - \rho(k) \cdot v(k)\right] - $$

$$-\frac{T}{\Delta}\left[\zeta_1(k) + \phi(k)\right]. \qquad (4.6)$$

Since we consider a single segment, eqn. (3.2) cannot be used as a model equation for mean speed v. Similarly to the treatment in /48/ we use the simple equation

$$v(k+1) = v(k) + \eta(k) \qquad (4.7)$$

for the mean speed of the vehicles being in the considered segment where $\eta(k)$ is a zero mean noise. During the kth time interval, however, there are q_o T vehicles entering the segment and q_1T vehicles leaving it. The mean speeds of these vehicles should also be considered in the specification of the overall mean speed, i.e.

$$v(k+1) = \frac{[\rho(k)\ \Delta - q_1(k)\ T]\ [v(k) + \eta(k)] + q_o(k)\ T\ w_o(k)}{\rho(k)\ \Delta + [q_o(k) - q_1(k)]\ T} \qquad (4.8)$$

Recall that $\rho(k)\cdot\Delta$ is the number of vehicles in the segment at time k T. Substituting eqns. (4.1)-(4.3) and (4.5) into (4.8) we get

$$v(k+1) = \left.\frac{\rho\ \Delta - (\rho v\ T - \phi\ T)(v+\eta) + (u_1 - \zeta_1)(u_2 - \zeta_3)\ T}{\rho\Delta + (u_1 - \rho v - \zeta_1 - \phi)}\right|_{(k)} . \qquad (4.9)$$

For the exit mean speed one can simply write

$$w_1(k) = v(k) + \zeta_5(k) \qquad (4.10)$$

with zero mean $\zeta_5(k)$. With eqn. (4.4) we get

$$y_2(k) = v(k) + \zeta_5(k) + \zeta_4(k) . \qquad (4.11)$$

Equations (4.6) and (4.8) have the general form (A1.16) and equations (4.5) and (4.11) have the general form (A1.17) required in Appendix A1. Hence, application of the Kalman filtering equations yields the suboptimal nonlinear estimator of eqn. (A1.20) with Kalman gain matrix K(k) given by eqns. (A1.18), (A1.19), (A1.13), (A1.15).

b) Results

The simultaneous recursive estimators were tested using data of a 3/4-mi segment of the Long Island Express way, New York. Photographs of the traffic were taken at 2-s intervals and the speed and trajectory of individual cars were established using successive picture frames. The above data were utilized to de rive the true values of density and mean speed at 2-s intervals. Also the data were used to generate speed and arrival time of cars crossing the sensor locations at the both ends of the segment. Initial choices have been taken inaccurate.

The estimator equations were implemented on a digital computer utilizing the sensor data from both segment ends. The output of the estimators was compared with the true segment density and mean speed obtained from the real data. Figure 4.4 shows the comparison of the true car count and its estimate. Figure 4.5 shows the plots of the true segment mean speed and its estimate.

The tracking capability of the estimator are clearly demonstrated in both figures. Unfortunately, the traffic data utilized do not exhibit large variations in speed and do not contain congestion data, which would represent a much more difficult estimation task.

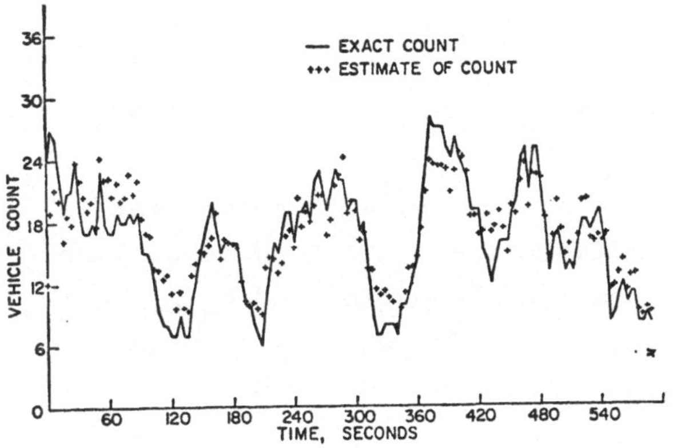

Figure 4.4 - Measured and estimated vehicle count.

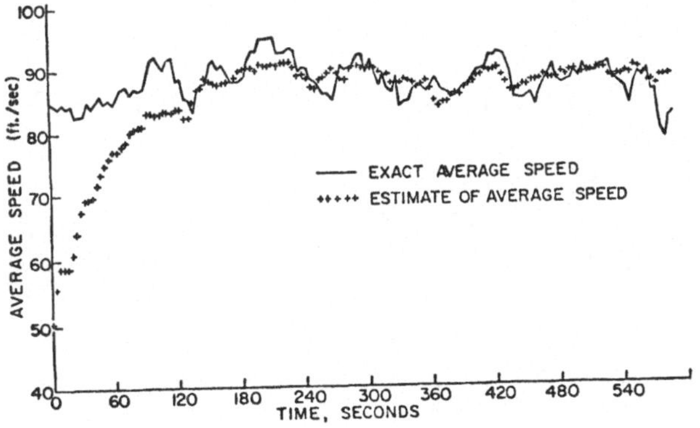

Figure 4.5 - Measured and estimated mean speed.

4.3 State estimation for a long freeway section

In view of the excellent identification results,it seems reasonable to consider a long section consisting of several freeway segments as the process under consideration. Compared to the treatment of section 4.2, this approach would reduce the presence detectors needed and hence the implementation cost by a considerable amount /44/.

a) Model equations

Consider the freeway section of figure 3.1.Following observations are available at time k

$$u_1(k) = q_o(k) + \zeta_1(k) \qquad (4.12)$$

$$y_1(k) = q_n(k) + \zeta_2(k) \qquad (4.13)$$

$$y_2(k) = w_o(k) + \zeta_3(k) \qquad (4.14)$$

$$y_3(k) = w_n(k) + \zeta_4(k) \qquad (4.15)$$

with zero mean noises $\zeta_i(k)$, $i = 1, \ldots, 4$, with known statistics. Eqns. (4.12) - (4.15) indicate that we intend to utilize $q_o(k)$ as an input variable in the model equation and $q_n(k)$, $w_o(k)$, $w_n(k)$ as output variables. Similarly as for eqn. (4.5), eqn. (4.13) yields

$$y_1(k) = \rho_n(k) \cdot v_n(k) + \phi(k) + \zeta_2(k). \qquad (4.16)$$

For $w_o(k)$, $w_n(k)$ we set

$$w_o(k) = v_1(k) + \eta_o(k) \qquad w_n(k) = v_n(k) + \eta_n(k) \qquad (4.17)$$

and we get from eqns. (4.14), (4.15)

$$y_2(k) = v_1(k) + \eta_o(k) + \zeta_3(k) \qquad (4.18)$$

$$y_3(k) = v_n(k) + \eta_n(k) + \zeta_4(k) \qquad (4.19)$$

where $\eta_o(k)$, $\eta_n(k)$ are additional zero mean noises. Traffic density for the first segment is derived by setting eqn. (4.12) into eqn. (3.1)

$$\rho_1(k+1) = \rho_1(k) + \frac{T}{\Delta_1} \left[u_1(k) - q_1(k)\right] - \zeta_1(k). \tag{4.20}$$

Traffic volumes between segments are given by eqn. (3.3) with an additional zero mean noise considering model inaccuracy

$$q_j(k) = \alpha \cdot \rho_j(k) \cdot v_j(k) + (1-\alpha)\rho_{j+1}(k)\ v_{j+1}(k) + \zeta_5(k) \tag{4.21}$$

$$j = 1, \ldots, n-1,$$

whereas for the last segment

$$q_n(K) = \rho_n(k) \cdot v_n(k) + \phi(k) \tag{4.22}$$

is assumed according to eqn. (4.16). Substituting of eqns. (4.21), (4.22) into eqn. (3.1) for j = 2, ..., n provides the model equations for the traffic density of the segments 2,...,n. Mean speed is given by eqn. (3.2) with an additional zero mean noise $\zeta_6(k)$. For the case j=1, $v_o=v_1$ is set, whilst for j=n, $\rho_{j+1}=\rho_j$ has been assumed. This completes the derivation of the model equations for Kalman filtering. Equations (3.1) and (3.2) with above modifications correspond to eqn. (A1.16) of Appendix 1 and equations (4.16), (4.18), (4.19) correspond to the measurement equation (A1.17), and a suboptimal extended Kalman Filter can be derived.

b) Results

The recursive estimator for a long section has been tested on the basis of real traffic data from the same freeway section as in section 3. Since no direct measures of traffic density and mean speed have been available, the comparison of estimated and real traffic data was performed by utilizing the internal measurement site of figure 3.1. The time interval was 10 sec.

The results for the same measurement set as in figure 3.3 are shown in figure 4.6. The breakdown of traffic after ca. 30 minutes is reproduced fairly accurately. Figure 4.7 gives the results for the measurement set of figure 3.5. The short congestion of about 10 minutes is reproduced by the estimator as well.

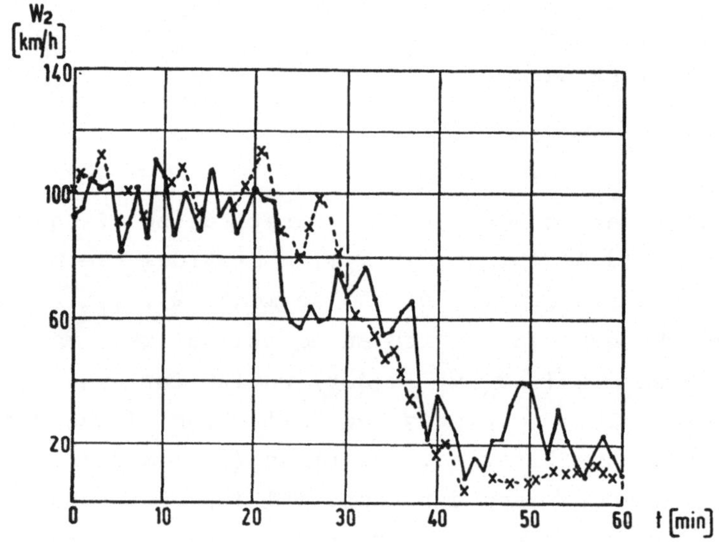

Figure 4.6 - Measured (——) and estimated (---) mean speed.

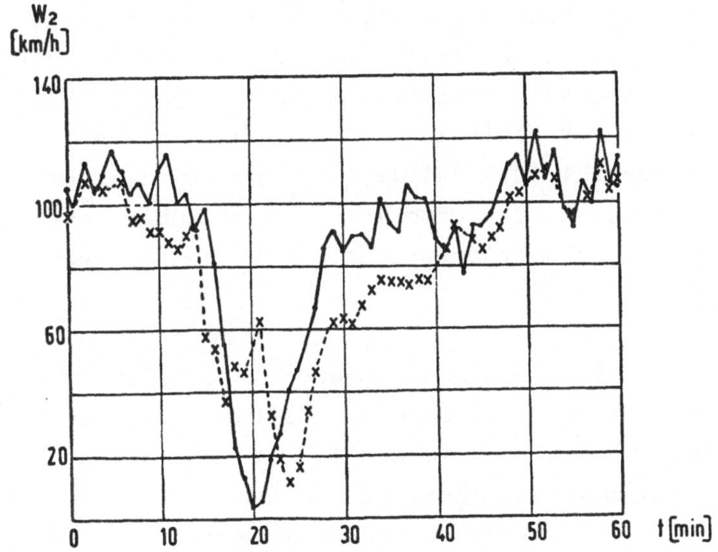

Figure 4.7 - Measured (——) and estimated (---) mean speed.

In order to investigate the tracking capabilities of the estimation algorithm, the case of uncongested real traffic with congested initial condition for the estimator has been considered. Results shown in figure 4.8 show that the estimated state reaches after 15 minutes the real traffic state in spite of the big initial deviation. In figure 4.8 the results obtained on the basis of the structure with four inputs (figure 4.1) are also shown. It is evident that without the correction mechanism provided by filtering techniques,results will be biased in the case of a wrong initial condition.

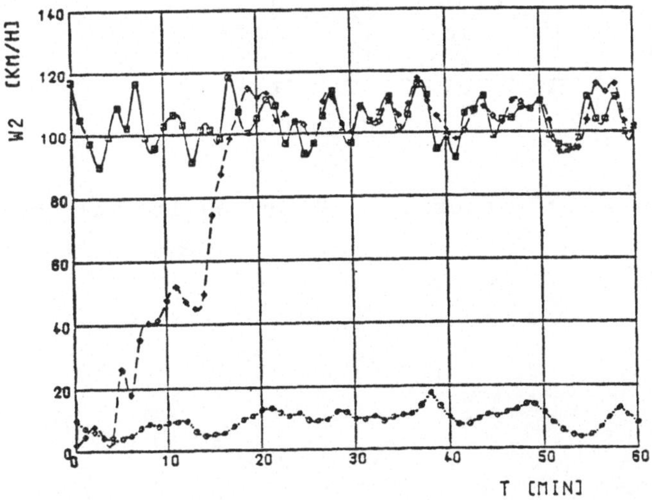

Figure 4.8 - Mean speed measure (——), estimated by an

In the opposite case, where congested real traffic state but uncongested initial consition is considered, the estimator fails to accurately track the real traffic behaviour, as shown in figure 4.9. This is a consequence of the fact that linearization (A1.18), (A1.19) are performed around an inadequate state point. Introduction of a switching strategy based on heuristic considerations led to an improvement of the estimates.

Finally, the case of biased measurements is considered in figure 4.10. Measures of $q_n(k)$ have been altered so that the passing of two vehicles each minute is not detected. Figure 4.10 shows that estimates with extended Kalman Filter are not considerably influenced whilst an algorithm without correction (figure 4.1) produces a non existing congestion.

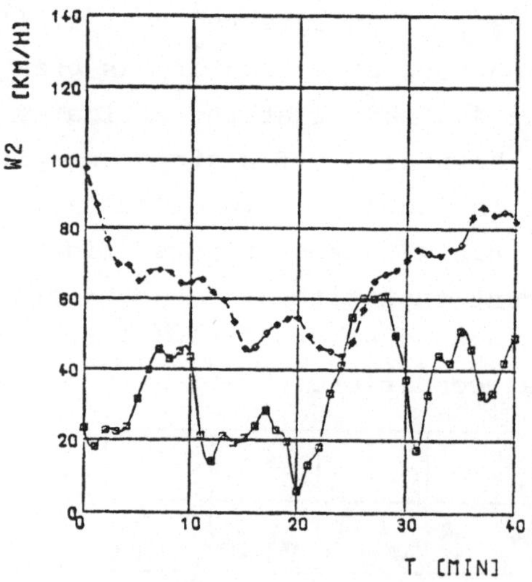

Figure 4.9 Mean speed measured (——) and estimated(---)

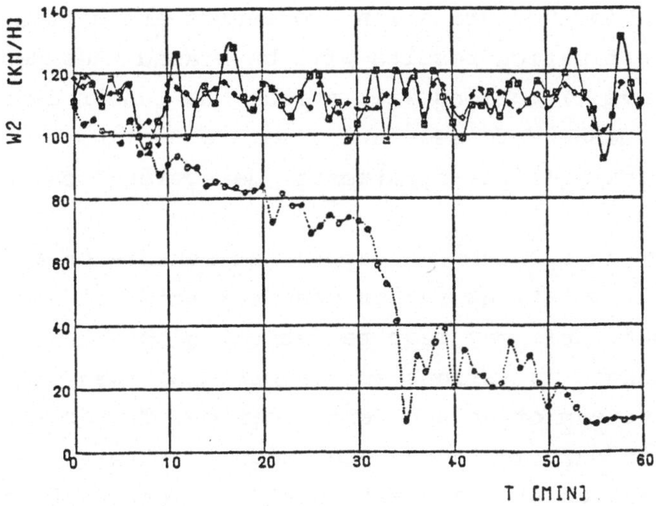

Figure 4.10 Mean speed measured (——), estimated with EKF (---), and estimated without correction (...)

4.4 Detection of incidents

Accurate detection of an incident or accident on a freeway when only local sensor data are available is one of the most challenging problems in the area of traffic data processing. From the traffic point of view, accidents may occur in many degrees of severity depending on whether any, one, or more lanes are completely or partially closed which corresponds to a significant change of the processes behaviour. In any case, models used in the estimation procedures are getting inaccurate and estimation results will be biased. Besides, each of the above situations will affect the sensor data in a particular manner. Consequently, we are dealing with a highly complex multiple-hypothesis detection problem.

The signature of an accident may appear at sensors on a "pattern" of signal variations rather than variation of any individual output. This may require detecting occurrence of a pattern of changes in various sensor outputs taken upstream and downstream of an accident location. To add to the complexitites, in many instances and in contrast with the usual signal-detection problem, extending the period of observation does not necessarily aid the detection performance, since if an accident is not detected soon, the traffic may readjust such that the signature at sensors may significantly diminish in magnitude. On the other hand, control actions will be significantly delayed.

Applications of several techniques for incident detection have been reported in the literature /55-59/. Pattern classification techniques seem to provide quite promising results but lead often to 'false alarms', i.e. detection of incidents which do not exist in reality. Reducing the false alarm rates leads to a reduction of the real incidents detected.

Recently, an alternative approach has been proposed by Willsky et. al. /57,128/ by using the Multiple Model (MM) method. The MM method for system identification has been considered by several researchers. The interested reader is referred to the references cites in /57,128/ for a detailed development of the technique.

The method addresses the problem of identifying a linear Gaussian system

$$\dot{\underline{x}}(t) = A\,\underline{x}(t) + \underline{w}(t) \tag{4.23}$$

$$\underline{z}(t_k) = C\,\underline{x}(t_k) + \underline{v}(t_k) \tag{4.24}$$

given the measurements $Z_k = \{z(t_1),\ldots, z(t_k)\}$ and a set of hypothesized models $(i = 1,\ldots,N)$

$$\dot{\underline{x}}_i = A_i\,\underline{x}_i(t) + \underline{w}_i(t) \tag{4.25}$$

$$\underline{z}(t_k) = C_i\underline{x}_i(t_k) + \underline{v}_i(t_k)\,. \tag{4.26}$$

The output of the MM method is the set $P_i(t_k)$ of conditional probabilities for the validity of each of the models given Z_k.A kalman filter is implemented for each of the N models and the measurement residuals

$$\underline{\gamma}_i(t_{k+1}) = \underline{z}(t_{k+1}) - C_i\hat{\underline{x}}_i(t_{k+1} \mid t_k) \tag{4.27}$$

from each filter are used to update the $p_i(t_k)$ according to suitably chosen equations.

The MM method has been adopted for use with the model D of section 2.2. A number of comments need to be made about this design and about the MM method in general.

i) A linearized Kalman filter has been used for each of the following hypotheses.
   a) For the normal model D in its continuous time version.
   b) For the model representing an incident on segment i,the dynamics (2.28) are modified by replacing the normal v_e curve, figure 2.3, on segment i with a reduced capacity curve.
   c) For the model representing a pulse of traffic on segment i, the dynamics (2.13) are modified by including an artificial input flow $r_i(t)$.

d) In addition to the above, there are also a set of models and associated filters representing sensor failures. Sensor failures have been modelled by modifying the measurement equation (4.26)

$$\underline{z}(t_k) = C'\underline{x}(t_k)+\underline{N}(t_k) \tag{4.28}$$

where C' is diagonal, with 1's along the diagonal except for a zero in the location corresponding to the particular state measurement which is hypothesized to be faulty.

ii) Densities and mean speeds in each segment are assumed as known measurements. Since they cannot be directly measured, an estimation algorithm like the ones described before must be also implemented.

iii) The results with respect to the conditional probabilities p_i are derived assuming that :

a) the actual system and all of the hypotheses are linear-Gaussian;

b) one of the hypotheses matches the true system;

and

c) the true system does not switch from one hypothesis to another (corresponding, for example, to the onset of an incident).

None of these assumptions is valid, and thus some comments are in order. Assumption 3a) essentially addresses the problem of the utility of the linearized Kalman filter, i.e., assuming the dynamic model is correct, is it valid to postulate that the filter residuals will be zero-mean, white, with precomputed covariance? The second assumption implied that (under assumption a)) the residuals from one of the filters will be white and zero mean. In practice this is never precisely the case, but experiments showed that neither of these assumptions has caused great problems. A number of explanations can be given to account for this, but there are no general results that predict when these filters will work well. However, the experience of the authors of /128/ has been that, while the estimates from the filters may be sensitive to linearization and model uncertainties, a discrete decision process based on the filter residuals should work well, as long as the models for the several hypotheses are sufficiently different. Intuitively, this can be thought of as a signal-to-noise ratio problem, where the effects of the assumptions add uncertainty. In this sense, assumptions a) and b) will limit the minimum size incident that can be detected, where size is to be interpreted as the magnitude of the effect of the incident on the dynamics. For example, one may be able to detect a stalled car, which causes severe and localized capacity reduction, but the smaller effect caused, say, by debris on the road may not be detectable. Furthermore, the effect of an incident increases in magnitude as the level of traffic increases. Thus, one might expect there to be a minimum flow level, such that it is impossible to detect incidents in traffic lighter than that level.

Assumption c) can lead to difficulties in the ability of MM to detect incidents as they occur, i.e., before the occurrence of an incident on segment i, the probability for this hypothesis may become so small that the system will not be able to respond quickly after the incident has occurred. The remedy employed in the work is a relatively common one - a lower bound is set on any probability. As we will see, this leads to good response characteristics. We note also that the Kalman filter based on a pulse of traffic on link i is unstable if no such pulse is there (the Kalman filter has a constant driving term in the ρ_i equation not present in the true system). Thus, if such a pulse were to develop at some point in time, the filter estimate for this hypothesis might already be so much in error that the MM system might not detect the pulse. To overcome this, whenever the probability of a pulse model falls below 0.05, the estimate produced by this filter is reset to the estimate for the most probable model.

Several simulation tests with macroscopic and microscopic freeway traffic models are reported in /128/. The reported results are encouraging:

a) Detection performance was uniformly good over the entire range of actual mean flows used (900-2000 veh/h/lane). No false alarms were observed, no inccorrect detections. Figure 4.11, illustrates a typical performance of the MM system.

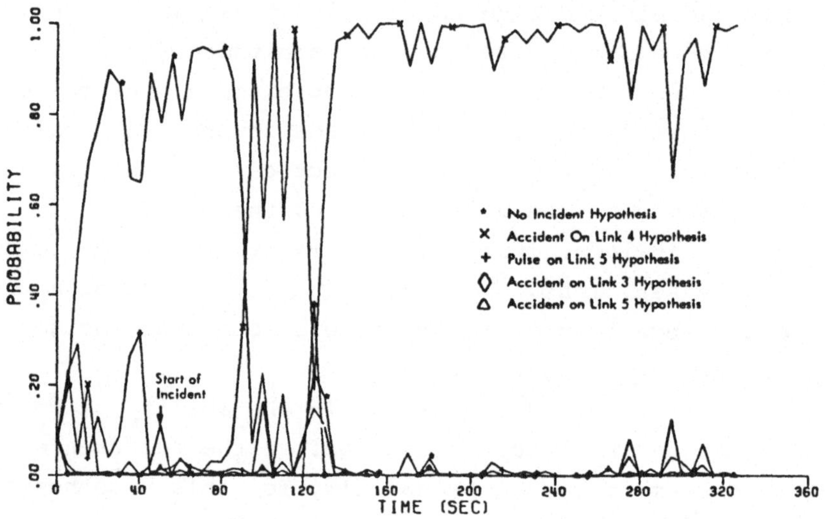

Figure 4.11 - MM probability plot. Accident on segment 4; macroscopic model simulation; nominal flow = 1000 veh/h/lane.

b) Performance is somewhat degraded when the actual measurement variances are a factor 16 larger than the nominal. All incidents were correctly identified with, however, increased detection delay.

c) Large initial estimation errors cause only transient effect on MM. Performance is excellent after the initial startup.

5. CONTROL PROBLEMS IN TRAFFIC DYNAMICS

The macroscopic models of freeway and road traffic presented in section 2 provide a basis for the development of control strategies through application of wellknown automatic control concepts. In this section we will formulate and solve some traffic control problems which have been considered in recent research works. We will distinguish again between control of freeway traffic and control of road traffic.

5.1 Freeway traffic control

As already mentioned in section 1, the main purpose of controlling freeway traffic is to prevent or eliminate the built-up of congestions in order to keep traffic flow in the freeway at high levels and increase traffic flow security. We will distinguish between two kinds of congestion.

Recurrent congestions occur due to high demand exceeding the limited capacity of freeways during the rush hours. In view of the results of section 2 it is easy to understand that the reason leading to nonrecurrent congestions are high density values ρ exceeding the critical density ρ_{cr}. Congestions caused by too high demand occur at specific known times-of-day or days-of-year.

Nonrecurrent congestions are due to unusual, unexpected circumstances such as accidents or other incidents affecting traffic conditions. For example, an incident can partially or totally blockage one or more freeway lanes. This corresponds to a significant reduction of freeway capacity and usually leads to the built up of congestion. Clearing the closed lanes is not sufficient for elimination of the built congestion, if no control actions are supplied.

a) Input variables

Let us first discuss possible ways of taking influence on the process of traffic flow.

a1) Ramp metering

A direct influence on the processes operation can be taken by metering the on-ramp volumes r in eqn. (2.13). This can be done by installing common traffic lights at the on-ramps of the freeway /60,61/. The underlying idea is to keep traffic density at values near the critical density (figure 2.3) guaranteeing traffic operation at maximum traffic volume. In case of non-recurrent congestion, on-ramp volumes should be reduced in order to enable a quick release.

Adjusting of on-ramp volumes must be performed subject to some constraints. On-ramp volumes cannot be higher than the current demand at a given entrance ramp, i.e.

$$r_i(k) \leq d_i(k) + l_i(k)/T \qquad (5.1)$$

where d_i is the arriving volume or demand volume and l_i is the queue length described by the difference equation

$$l_i(k+1) = l_i(k) + T\,[d_i(k) - r_i(k)]. \qquad (5.2)$$

Clearly, an entrance queue will be formed any time the volume of traffic permitted to enter the freeway is less than the volume desiring to use the ramp.

Due to the geometric characteristics of the entrance ramp, there is a maximum on-ramp volume value $r_{i,max}$ which can not be exceeded and hence

$$r_i(k) \leq r_{i,max} \qquad (5.3)$$

must always hold. On the other hand, too low rates of on-ramp volumes lead waiting drivers to judge the metering signal to be malfunctioning /10,62/. Hence a reasonable lower limit should be posed

$$r_i(k) \geq r_{i,min} > 0\,. \qquad (5.4)$$

Finally, metering on-ramp volumes can lead to long entrance queues during the rush hours. In order to avoid collisions of traffic in surface streets, a maximum queue length $l_{i,max}$ should not be exceeded, i.e.

$$l_i(k) \leq l_{i,max} . \qquad (5.5)$$

Substituting eqn. (5.2) into (5.5) we obtain

$$r_i(k) \geq d_i(k) - \frac{1}{T} \left[l_{i,max} - l_i(k)\right]. \qquad (5.6)$$

Summarizing equations (5.1), (5.3), (5.4), (5.6), the admissible control region for on-ramp volumes is given by

$$\max \{\underline{r}_{min}, \underline{d}(k) - \frac{1}{T} [\underline{l}_{max} - \underline{l}(k)]\} \leq \underline{r}(k) \leq$$

$$\leq \min \{\underline{r}_{max}, \underline{d}(k) + \frac{1}{T} \underline{l}(k) \}. \qquad (5.7)$$

Off-ramp volumes s_i enter eqn. (2.13) in the same way as on-ramp volumes. Since metering of off-ramp volumes is not practicable, they cannot be considered as input variables. Nontheless, variable information signs installed along the freeway and warning vehicle drivers about the possible existence of congestion at a downstream location, might motivate short-trip-drivers to leave the freeway. This corresponds to an indirect off-ramp volume setting.

a2) Variable message signs

Extensive observations of freeway traffic phenomena under influence of variable me ssages provided by traffic signs led to fairly interesting results. In particular, experiments with variable speed limitation signs have been performed in European countries /12, 63; 64/. It has been found that reasonable utilization of speed limitations during rush hours leads to an increase of capacity and stability of traffic flow. This corresponds to an increase of q_{max} and ρ_{cr} in figure 2.3. Based on the results of /64/, an analytical formula has been proposed in /51/ extending the speed-density relationship (3.4) so as to include the impact of speed limitations :

$$v_e(b,\rho) = v_f \cdot b \left[1 - \left(\frac{\rho}{\rho_{max}}\right)^{m(3-2b)}\right]^{l} . \qquad (5.8)$$

b is an input variable corresponding to speed limitation values. b = 1 corresponds to no speed limitation as is easily verified by eqns. (3.4), (5.8). Decreasing values of b correspond to particular speed limitation values according to a table given in /51/.

The role of speed limitations b is best demonstrated on the basis of the steady-state volume density characteristic corresponding to eqn. (5.8) and shown in figure 5.1 for various b-values. ρ_{cr} increases monotonically with decreasing b. q_{max} achieves a maximum value for $b \approx 0.75$. From figure 5.1 one can see that speed limitation leads to an increase of traffic volume for high densities only. The admissible control region for speed limitations is given by

$$0.7 \leq b_i \leq 1 . \qquad (5.9)$$

Besides, only discrete speed limitation values are admissible, i.e.

$$b_i \varepsilon \{ 0.7, \ 0.8, \ 0.8 \ , \ 1\} . \qquad (5.10)$$

The frequency of speed limitation changes should be low, in order to avoid driver's irritation

$$b_i(k) = b_i(k+1) = \ldots = b_i(k+8) \ , \quad k = 0(8)K. \qquad (5.11)$$

For a time interval T = 15s, eqn. (5.11) permits only one speed

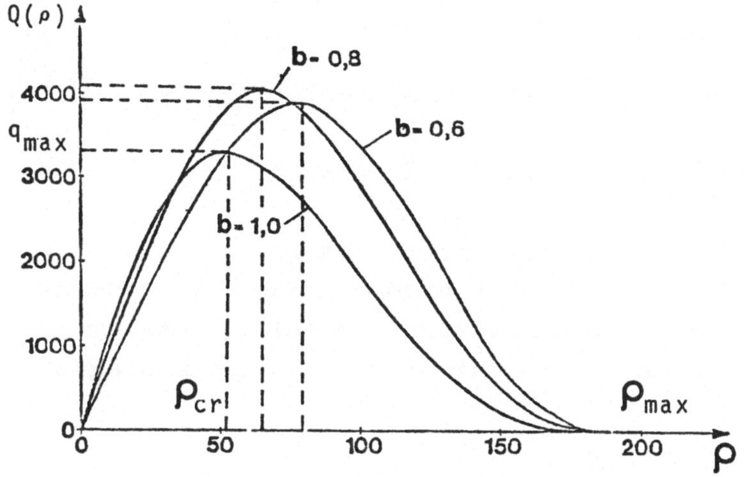

Figure 5.1 - Volume-density characteristic with speed limitations.

limitation change every 2 minutes.

b) Control objective

Several control objectives for freeway traffic have been proposed by various researchers /10, 65, 66/. Minimization of the total time T_s spent by all drivers on the freeway system seems to be the most reasonable one. Total time spent on the freeway includes total travel time and total waiting time at the on-ramps. Minimization of total time spent on the freeway implies minimization of delays caused by congestions and is thus a suitable requirement in order to prevent or eliminate both recurrent and nonrecurrent congestions.
T_s is given by the sum

$$T_s = T_T + T_W \tag{5.12}$$

where T_T is the total travel time and T_W is the total waiting time. The total travel time for a given time horizon K T is given by the sum

$$T_T = \sum_{k=0}^{K} T \cdot \underline{\rho}(k)^T \underline{\Delta} \, . \tag{5.13}$$

The total waiting time for the same time horizon is

$$T_W = \sum_{k=0}^{K} T \cdot \underline{l}\,(k)^T \underline{e} \tag{5.14}$$

with $\underline{e}$ the unit vector of appropriate dimension. Minimizing the total time T_s is equivalent to maximizing a weighted sum of the total freeway output /67/. To see this, introduce a variable $N(k)$ denoting the number of cars in the freeway system. Obviously

$$N(k) = \underline{\rho}(K)^T \underline{\Delta} + \underline{l}(k)^T \underline{e} \, . \tag{5.15}$$

On the other hand, the following conservation equation holds

$$N(k+1) = N(k) + T \cdot [\underline{d}(k) - \underline{s}(k)]^T \underline{e} \tag{5.16}$$

which implies that

$$N(k) = N(0) + \sum_{\varkappa=0}^{K-1} T \cdot [\, \underline{d}(\varkappa) - \underline{s}(\varkappa) \,]^T \underline{e} \, . \tag{5.17}$$

Combining eqns. (5.12) - (5.15) and (5.17) we obtain

$$T_s = T \cdot \sum_{k=1}^{K} \left\{ N(0) + T \cdot \sum_{\varkappa=0}^{k-1} [\underline{d}(\varkappa) - \underline{s}(\varkappa)]^T \underline{e} \right\} + N(0) \tag{5.18}$$

and since $N(0)$ and $\underline{d}(k)$ cannot be influenced by the control system

$$(T_s \rightarrow \min) \Leftrightarrow (S = \sum_{k=1}^{K} \sum_{\varkappa=0}^{k-1} \underline{s}(\varkappa)^T \underline{e} \rightarrow \max) \tag{5.19}$$

where the quantity on the right hand side of (5.19) is equal to

$$S = \sum_{k=0}^{K} (K-k) \cdot \underline{s}(k)^T \underline{e} \qquad (5.20)$$

which is a weighted sum of the freeway output.

Minimization of the total travel time is formally achieved by letting $\underline{\rho}(k) = \underline{0}$ in eqn. (5.13), which corresponds to an on-ramp closure. Obviously, this represents the worst possible solution, since the waiting time T_W and hence T_S will grow to a maximum. On the other hand, minimization of the total waiting time is achieved for $\underline{l}(k) = \underline{0}$, which corresponds to the uncontrolled case $\underline{r}(k) = \underline{d}(k)$. In this case, congestions will be built up leading to a substantial raise of total travel time T_T and T_S, if demand exceeds freeway capacity. Thus, in distinction to these extreme cases, fulfilment of objective (5.19) is expected to lead to a more or less slight raise of the waiting time T_W and a substantial decrease of the travel time T_T so that the sum of both, i.e. the total time spent on the freeway, is minimized.

c) Control on the basis of model A: steady-state time-of-day control

By inspection of past traffic data it can be seen that demand volumes d_i and O-D rates α_{ij} are slowly varying compared to traffic dynamics and can hence be assumed constant over time periods of 15-30 minutes /10,68/. This time period is long enough for the traffic flow on short freeways to reach a steady-state. Specification of such an optimal steady-state is the objective of the control problems described in this section.

The values of demand volumes d_i and α_{ij} at several time periods during the rush hours can be anticipated according to a prediction scheme distinguishing working-days from weekend-days. A typical example of a demand volume during a week is shown in figure 5.2 /69/. It can be seen, that conditions at each working day are almost identical so that the same fixed time control algorithms can be applied for all working days.

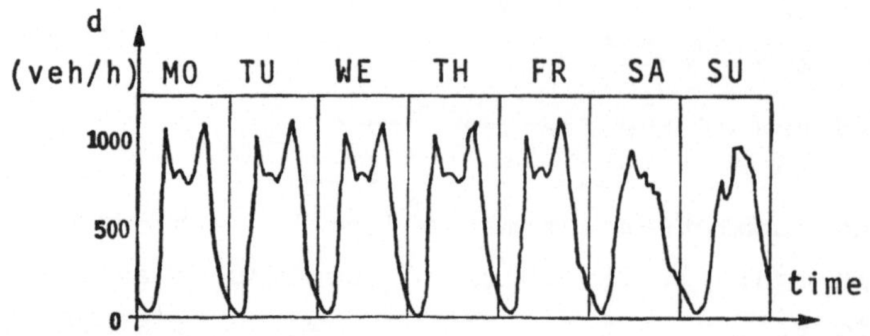

Figure 5.2 - Typical demand volume from a german freeway.

The control strategy should only depend upon the time-of-day, since demand volumes and O-D rates are slowly varying during the day.

Since we are interested only on the steady-state condition, eqn. (2.15) is considered as the traffic model. In order to keep traffic flow on the left hand side of the volume-density characteristic, we require

$$\underline{\bar{q}} \leqslant \underline{q}_k \tag{5.21}$$

where $q_{k,i}$ are the sections' capacities. For example, we can set $\underline{q}_k = \underline{q}_{max}$, the maximum traffic volume. Introducing the O-D matrix $\underline{A}$ and substituting eqn. (2.15) into eqn. (5.21) we obtain

$$\underline{A}\underline{\bar{r}} \leqslant \underline{q}_k \, . \tag{5.22}$$

In the steady-state case, inequality constraints (5.7) become

$$\max \{\underline{r}_{min}, \underline{d} - \frac{1}{T} [\underline{l}_{max} - \underline{l}(0)] \leqslant \underline{r} \leqslant$$

$$\leqslant \min \{\underline{r}_{max}, \underline{d} + \frac{1}{T} \underline{l}(0)\} \tag{5.23}$$

where T denotes the length of the time-of-day period. The steady-state version of performance criterion (5.20) is

$$\bar{J} = \bar{\underline{s}}^T \underline{e} \rightarrow \max. \tag{5.24}$$

But in the steady-state, the total freeway output expressed by eqn. (5.24) equals the total freeway input and hence

$$\bar{J} = \bar{\underline{r}}^T \underline{e} \rightarrow \max \tag{5.25}$$

is equivalent to (5.24).

The steady-state time-of- day control problem P1 can now be formulated /70/:

Given $\bar{\underline{d}}$, $\bar{A}$, $\underline{q}_k$
Find $\bar{\underline{r}}$
so as to minimize $\bar{J}$, eqn. (5.25)
subject to eqns. (5.22), (5.23).

It can be easily verified that P1 constitutes a linear programming problem which can be solved by use of wellknown algorithms. Speed limitations are not included in P1. In order to do that, the following procedure should be evaluated:

1. Set $q_{k,i} = 4000$ which corresponds to $b_i = 0.8$ (maximum capacity value according to figure 5.1).
2. Solve P1.
3. Set speed limitations according to

$$b_i = \begin{cases} 1 & \text{if} \quad \bar{q}_i \leq 3300 \\ 0.9 & \text{if} \quad 3300 \leq \bar{q}_i \leq 3600 \\ 0.8 & \text{if} \quad 3600 \leq \bar{q}_i \leq 4000 \end{cases} \tag{5.26}$$

according to figure 5.1.

Because of their relative simplicity, time-of-day control plans have usually been the first attempted in a freeway control system /71,72/. Of course, time-of-day control strategies can only be applied in order to prevent recurrent congestion.

Nonreccurent congestions caused by freeway incidents cannot be eliminated by time-of-day control, which is an open-loop control scheme ignoring the current traffic conditions. Alternative formulations of steady-state time-of-day control problems can be found in /73-78/.

As an illustration of the approach, a numerical example will be presented next. Consider a hypothetical freeway with seven on-ramps and seven off-ramps. Demand volumes at the on-ramps are given by

$$\bar{d}_1=2000,\ \bar{d}_2 = \bar{d}_3 = \bar{d}_4 = \bar{d}_5 = 600\ ,\quad \bar{d}_6 = \bar{d}_7 = 400 \qquad (5.27)$$

and the O-D matrix

$$\bar{A} = \begin{bmatrix} 1 & 0.95 & 0.9 & 0.85 & 0.8 & 0.75 & 0.7 \\ & 1 & 0.98 & 0.95 & 0.9 & 0.8 & 0.76 \\ & & 1 & 0.98 & 0.95 & 0.92 & 0.82 \\ & & & 1 & 0.98 & 0.92 & 0.8 \\ & & 0 & & 1 & 0.95 & 0.8 \\ & & & & & 1 & 0.95 \\ & & & & & & 1 \end{bmatrix}. \qquad (5.28)$$

The minimum admissible on-ramp volume values are :

$$r_{1,min} = 1500,\ r_{i,min} = 180\ ,\ i = 2, \ldots, 7. \qquad (5.29)$$

For simplicity, no maximum on-ramp volumes and no maximum queue lengths are considered. All section capacities are set to $q_{k,i} = 3100$. Speed limitations are not considered. Solution of the problem P1 yields

$$\bar{r}_1 = 2000,\ \bar{r}_2 = 600\ ,\ \bar{r}_3 = 180,\ \bar{r}_4 = 600\ ,\ \bar{r}_5 = 201,$$
$$\bar{r}_6 = 211\ ,\ \bar{r}_7 = 255. \qquad (5.30)$$

Obviously, the freeway demand exceeds the freeway capacity. Application of time-of-day control leads to the built up of queues at the on-ramps 3, 5, 6, 7. The queues might be released during the next time-of-day periods, during which they should be considered as additional demand.

d) Control on the basis of model C: dynamic time-of-day control

If we consider the whole time horizon K T, problem P1 is supposed to be solved independently K times. As already mentioned, the solution of problem P1 produces optimal input volumes in the steady-state. In long freeways, however, time intervalls of 15 or even 30 minutes are not long enough for the system to reach its steady-state condition. This can be illustrated by the following example.

Let us assume 80 km/h as a mean speed on a 40 km long freeway. If the input volumes are changed at the beginning of a new time period according to a fixed time-of-day control strategy, it will take approximately 30 min for the traffic flow to reach the new steady-state. However, if the freeway to be controlled is longer than 40 km, the steady-state can never be reached. Under such conditions repeated solution of P1 and application of the derived strategy cannot be expected to give optimal control for the real system.

An alternative way to derive improved time-of-day control strategies is by use of the linear but dynamic freeway model C (section 2.2.c). This approach results to the formulation of a single dunamic optimization problem P2 considering the whole interesting time horizon K T. The number of variables and constraints involved is of course accordingly higher but the problem is still a linear programming one. Details are given in /30/. Application of dynamic time-of-day control is capable of preventing recurrent congestion. Since no feedback paths are included, elimination of nonrecurrent congestion which requires traffic responsive control schemes cannot be obtained.

In order to illustrate the difference between the two time-of-day approaches, an example reported in /30/ will now be presented. Consider a hypothetical seven-section, two-lane freeway with seven on-ramps and seven off-ramps for a time horizon of three hours divided into nine 20-min long time periods (K=9). It is assumed that demands and origin-destination rates change their values after each time period. The values of the demands are given in table 5.1. Origin-destination rates take values similar to those of eqn. (5.28). A rapid increase of the demand in all ramps is assumed in the first periods, whilst in the last four time periods demand is again reduced to become undercritical. Following section lengths are taken (in km).

$\delta_1 = 4.2$, $\delta_2 = 4.6$, $\delta_3 = 8.2$, $\delta_4 = 8.0$, $\delta_5 = 4.4$, $\delta_6 = 4.4$, $\delta_7 = 9$

In order to calculate the corresponding mean travel times τ_j, a mean speed of 90 km/h was assumed. The minimum admissible

Period \ Section	1	2	3	4	5	6	7
1	1500	300	300	300	300	300	300
2	2000	600	600	600	600	400	600
3	3000	700	700	700	700	700	700
4	2800	600	600	600	600	500	500
5	2000	300	300	400	500	500	500
6	1500	200	200	200	300	300	200
7	1300	100	100	100	100	100	100
8	800	100	100	100	100	100	100
9	800	100	100	100	100	100	100

Table 5.1 - Demands $d_i(k)$, $i=1,\ldots,7$; $k = 1,\ldots,9$.

on-ramp volumes are given in eqn. (5.29). For simplicity no limitation of maximum queue length is considered in this example. Sections' capacities are set to $q_{k,i} = 3100$. Speed limitation are not considered.

Results obtained through K independent solutions of problem P1 for the K considered time periods are shown in table 5.2a) Results obtained through solution of a single optimization problem for the whole time horizon are given in table 5.2.b). In or-

Period \ Section	1	2	3	4	5	6	7
1	1500	300	300	300	300	300	300
2	2000	600	180	600	201	211	255
3	3000	192	180	180	180	194	234
4	2800	580	180	590	592	593	800
5	2000	800	446	180	180	213	200
6	1500	228	800	767	180	180	197
7	1300	100	714	283	800	423	181
8	800	100	100	100	767	785	800
9	800	100	100	100	100	100	133

a

Period \ Section	1	2	3	4	5	6	7
1	1500	300	300	300	300	300	300
2	1893	408	200	300	508	347	600
3	1871	408	200	425	383	347	325
4	2495	653	306	652	606	432	325
5	2324	344	582	203	180	189	840
6	2308	344	582	203	180	189	211
7	1709	344	630	213	175	201	211
8	800	100	100	705	869	595	190
9	800	100	100	100	100	400	100

b

Table 5.2 - Time-of-day control results
a) steady-state
b) dynamic

der to test and compare the efficacy of the control strategies, evolution of traffic flow on the hypothetical freeway using values of the input volumes provided by optimization problem P1 and P2 has been simulated by two different freeway traffic flow models. For comparison purposes also the uncontrolled case has been included in the simulation examples. For the uncontrolled case, on-ramp volumes are set equal to the corresponding demands. First the linear traffic model C is used in order to evaluate freeway traffic volumes. Figure 5.3 shows the traffic volume in freeway section 7 for the cases: a) without control, b) control obtained through solution of P1, c) control obtained through solution of P2. Clearly, the linear model C cannot reproduce a bottleneck situation. What it can show, however, is where and how much freeway capacity is excëeded, that is, at which locations traffic density will grow over its critical value. At such locations a bottleneck is to be expected.

In figure 5.3a) freeway capacity is violated and occurence of congestion is unavoidable. Freeway volumes are significantly reduced through steady-state ramp metering control but freeway capacity is still exceeded, (Figure 5.3b), so that a bottleneck will be a very probable consequence. Figure 5.3c) shows that dynamic time-of-day control exaclty holds the capacity restriction neither overloading nor underloading the freeway.

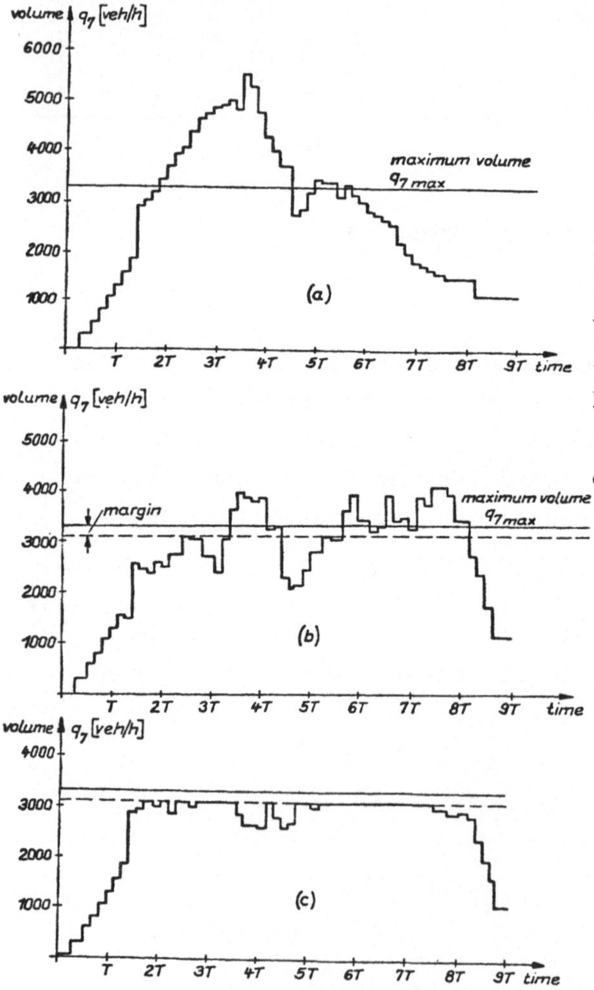

Figure 5.3 - Traffic volume in the 7th section simulated with model C in the cases
a) without control
b) steady-state time-of-day control
c) dynamic time-of-day control.

In order to show that above qualitative judgement is valid, traffic flow on the same hypothetical freeway was simulated with the much more accurate nonlinear model D. Figure 5.4 shows the evolution of traffic density for the three cases considered before. In the first two cases bottlenecks occur at locations, where traffic density exceeds the critical density value. This leads to a considerable reduction of mean speed and freeway traffic volume and consequently to an increase of the total travel time T_T in the freeway system. On the other hand, underloading of the freeway also occuring in case b) increases unnecessarily the total waiting time T_W. In the third case, however, traffic density remains in the left neighborhood of its critical value for the whole time of control operation, so that high performance and availability of the freeway is provided. The total times spent in the freeway for the three cases are (in veh.hours)

$$T_s^a = 6790, \quad T_s^b = 6010, \quad T_s^c = 5160.$$

This indicates the superiority of dynamic time-of-day control in preventing recurrent congestion.

e) Control on the basis of model D

Time-of-day control is not capable of eliminating nonrecurrent congestion occuring due to an unexpected incident on the freeway. In order to enable this, the nonlinear dynamics of traffic process should be taken into account. For this purpose, the highly accurate traffic flow model D can be utilized. Predicted demand and origin-destination rate trajectories are assumed available. They will be denoted by $\underline{d}_N(k)$ and $A_N(k)$. The following optimal control problem P3 can be formulated:

Given the predicted trajectories $\underline{d}_N(k)$, $A_N(k)$ and the initial condition for $\underline{\rho}(o)$, $\underline{v}(o)$, $\underline{l}(o)$

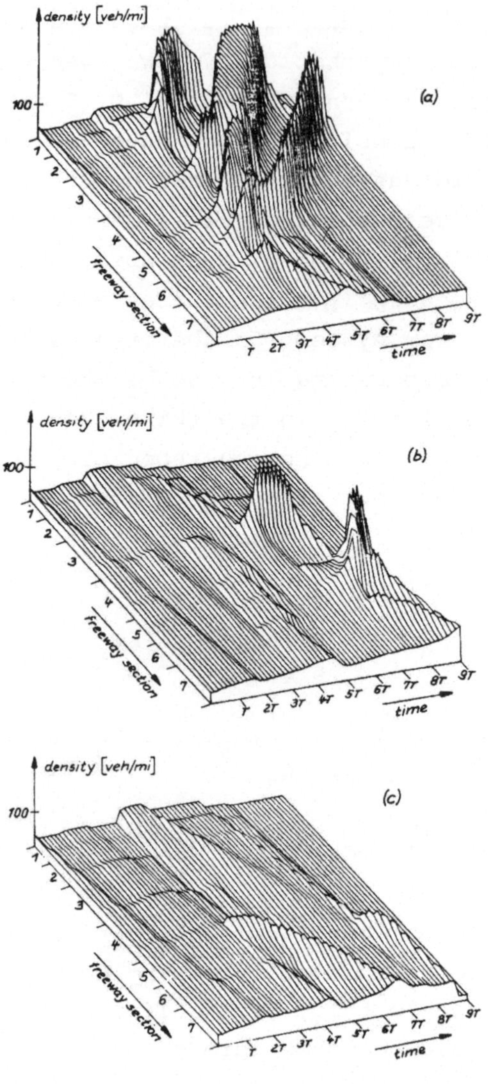

Figure 5.4 - Simulation of traffic flow with model D in the cases:

a) without control

b) steady-state time-of-day control

c) dynamic time-of-day control

Find $r^*(k)$, $b^*(k)$, $k = 0, \ldots, K-1$,
so as to minimize the total time spent in the freeway system given by eqns. (5.12) - (5.14)
subject to eqns. (3.1)[1] - (3.3) taken for each freeway segment and eqns. (5.2), (5.7) - (5.11) taken for each freeway section.

Optimization problem P3 can be solved by use of the Pontryagin's Minimum Principle /79/ (see also Appendix 2 for a short presentation). However, there are some difficulties connected with this approach. First, solution of the Two-Point-Boundary-Value-Problem (TPBVP) resulting from the necessary conditions for optimality, is an extremely difficult task because of the high order and the strong nonlinearities of the model used. Besides, singular subarcs are expected to occur, since input volumes $r(k)$ enter linearly the Hamiltonian. Last but not least, an open-loop control scheme such as the one represented by problem P3 is rather sensitive to changeable disturbance conditions. A detailled discussion of these problems and possible solution structures will be given in section 7.

5.2 Road traffic control

a) Input variables

Consider the road intersection described in section 2.3.a). Eqns. (2.38), (2.39) or (2.41) - (2.44) make obvious that the only input variable available is the green phase rate, resp. the service rate in one of the two directions which has been denoted by r. For practical reasons, the service rate should

(1) Eqn. (3.1) is considered in its complete form including on-ramp and off-ramp volumes, see also eqn. (2.13).

vary between an upper and a lower bound. This is so because a very short green phase is obviously wasteful and a very long red phase psychologically devastating. In fact, it is generally believed that a red phase exceeding a certain maximum leads some motorists to the belief that the light has failed and hence must be ignored. Thus, an admissible control region is defined by the inequalities

$$0 < r_{min} \leq r \leq r_{max} \; . \tag{5.31}$$

In order to avoid collisions of traffic between subsequent intersections, some maximum queue lengths should not be exceeded

$$\underline{l} \leq \underline{l}_{max} \; . \tag{5.32}$$

Besides, negative queue lengths should be excluded

$$\underline{l} \geq \underline{0} \; . \tag{5.33}$$

b) Control objective

For similar reasons as in freeway traffic (see section 5.1.b), minimization of total time spent in a road network should be the objective of the control system. In distinction to freeway traffic, the main interesting quantity in this context is the total waiting time, since total travel time can be regarded as constant and cannot be influenced by the green phase setting. In continuous time terms, the total waiting time is given by

$$T_w = \int_0^T \underline{l}^T \underline{e} \quad dt \tag{5.34}$$

where T is the time horizon of control operation. A somehow more "democratic" objective, leading to more or less equally long queues in all traffic directions is the quadratic one

$$F = \int_0^T ||\underline{l}||_Q^2 \, dt \qquad (5.35)$$

with a weighting matrix $Q \geq 0$, leading to a constraint linear quadratic problem.

At the end of the control operation,elimination of the queues could be required

$$\underline{l}(T) = \underline{0} \, . \qquad (5.36)$$

In that case, time horizon T could be viewed as being free.

c) Control of a single intersection

Following the treatment of /33, 34/, we will consider in this section the control of the simple intersection described by eqns. (2.38), (2.39). Control of more complex intersections will be shown to be very similar to the one of simple intersections. Control of road networks will be considered in section 6.2. The optimal control problem P4 for a simple intersection is the following :

Given the trajectoreis $d_1(t)$, $d_2(t)$, $t \varepsilon [0,T]$ and the initial condition $\underline{l}(0) = \underline{l}_0$

Find $r^*(t)$, $t \varepsilon [0,T]$ and T

so as to minimize the total waiting time, eqn. (5.34), subject to eqns. (2.38), (2.39), (5.31) -(5.33), (5.36).

Consideration of state variable constraints like the ones of eqns. (5.32), (5.33) is a very difficult task. For this reason we will ignore them in the problem solution and we will present some heuristic ways of considering them after the solution of the optimization problem.

The Hamiltonian of the problem is given by (see Appendix 2)

$$H = l_1 + l_2 + \lambda_1(a_1 - r) + \lambda_2(a_2 + b_2\, r) \tag{5.37}$$

where $a_1(t) = d_1(t)$

$$a_2(t) = d_2(t) - s_2(1 - \frac{L}{t_c})$$

$$b_2 = \frac{s_2}{s_1}$$

have been introduced for a shorter presentation. The Hamiltonian can be written in the form

$$H = H_1 + H_2 \cdot r \tag{5.38}$$

where H_1 is independent of r and H_2 is given by

$$H_2 = \lambda_2 b_2 - \lambda_1 . \tag{5.39}$$

Obviously, minimization of H, resp. H_2, with respect to r requires the following control rule

$$r^* = \begin{cases} r_{max} & \text{for} \quad H_2 < 0 \\ r_{min} & \text{for} \quad H_2 > 0 \\ \text{unknown} & \text{for} \quad H_2 \equiv 0 \end{cases} \tag{5.40}$$

The co state differential equations are given by

$$\dot{\lambda}_i = - \frac{\partial H}{\partial x_i} = - 1 \qquad , i = 1,2 \tag{5.41}$$

and hence

$$\lambda_i(t) = -t + c_i, \quad i = 1,2 \tag{5.42}$$

where c_i are integration constants. Substituting eqn. (5.42) into (5.39) we get

$$H_2 = t\ (1 - b_2) + b_2c_2 - c_1\ . \qquad (5.43)$$

Without loss of generality we assume $s_1 > s_2$, i.e. $b_2 < 1$, which excludes the singular case in eqn. (5.40). Since the coefficient of t is positive in eqn. (5.43), H_2 is an increasing linear function in time and hence it can only change from negative to positive. Consequently, in order to minimize H, r may only change from r_{max} to r_{min}. The possible swith-over time from r_{max} to r_{min} is given by the zero of H_2, namely,

$$t_s = (c_1 - b_2c_2)/(1-b_2). \qquad (5.44)$$

In order to specify the swith-over time we integrate eqns. (2.38), (2.39) assuming

$$r = \begin{cases} r_{max} & \text{for} \quad 0 \leqslant t \leqslant t_s \\ r_{min} & \text{for} \quad t_s \leqslant t \leqslant T\ . \end{cases} \qquad (5.45)$$

In order to get easily analytical results, we will now assume, that arrival rates d_1, d_2 are constant during the whole period of control operation, so that a_1, a_2 are constant as well. Notice, that the time evolution of d_1, d_2 has no influence on the main result represented by eqn. (5.45) and that it only influences the swith-over time t_s. Integration of (2.38), (2.39) for constant a_1, a_2 over $[0, t_s]$ yields

$$l_1(t_s) = (a_1 - r_{max})\ t_s + l_{10} \qquad (5.46)$$

$$l_2(t_s) = (a_2 + b_2r_{max})\ t_s + l_{20}\ . \qquad (5.47)$$

Integration over $[t_s, T]$ together with eqns. (5.36), (5.46), (5.47) yields

$$l_1(T) = (r_{min} - r_{max})\, t_s + l_{10} + (a_1 - r_{min})\, T = 0 \quad (5.48)$$

$$l_2(T) = (r_{max} - r_{min})\, t_s\, b_2 + l_{20} + (a_2 + b_2 r_{min})\, T = 0 \quad (5.49)$$

Eqns. (5.48), (5.49) constitute a linear system of two equations with two unknowns, t_s and T. Solving the system, we obtain

$$T = (-b_2 l_{10} + l_{20}) \,/\, (b_2 a_1 + a_2) \quad (5.50)$$

$$t_s = [(l_{10} a_2 - l_{20} a_1) + r_{min}(b_2 l_{10} + l_{20})] \,/\, [(b_2 a_1 + a_2)(r_{max} - r_{min})]. \quad (5.51)$$

Negative values for t_s, T might be the consequence of an unfeasible problem formulation.

Results are illustrated in figure 5.5 showing a possible control treatment. It becomes evident that the control system tries to minimize the delays by giving priority to the traffic stream with the greater saturation flow, which is by definition stream 1. In /33/ it has been shown that a further reduction of the total

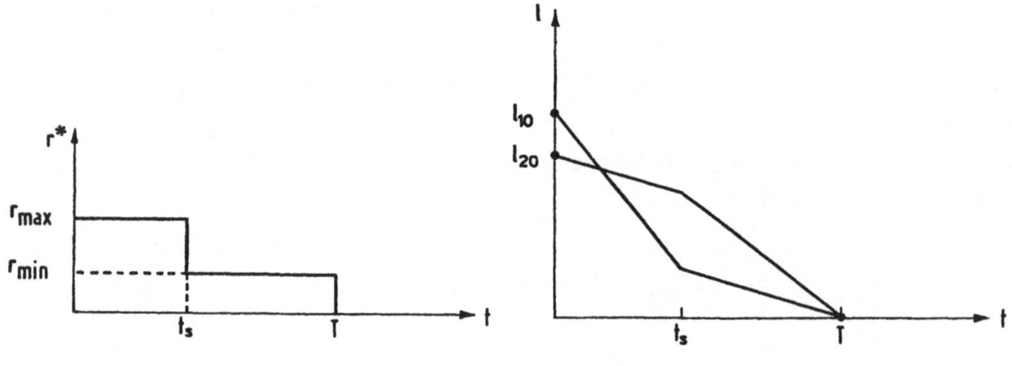

Figure 5.5 - Optimal control variable and queue lengths.

waiting time may be accomplished if we specify a fixed time horizon T' which is longer than T specified by eqn. (5.50) which means that we relax the requirement that both queues be dissolved at the same time, which is the earliest possible time. A further profitable trade-off of waiting times may then be accomplished by extending the maximum service rate for direction 1 past the switch-over point t_s determined above.

Above solution could lead to negative queues, if $r_{min} < a_1$ as illustrated in figure 5.6. Hence, care should be exercised in that case to assure that at no time does any queue become negative. If $r_{min} < a_2$, $l_{2,max}$ could be exceeded by application of the above solution. The switch-over time should be accordingly shortened in that case.

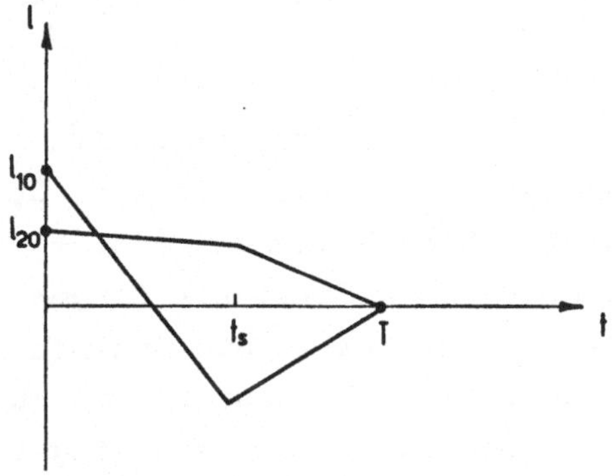

Figure 5.6
Possible queue evolution
for $r_{min} < a_1$

Let us now consider the case of a complex intersection as the one described by eqns. (2.41) - (2.44). Following a similar treatment, it is easy to verify that the main result, eqn. (5.45), is also valid for a complex intersection provided

$$s_{11} + s_{12} > s_{21} + s_{22} \tag{5.52}$$

which corresponds to the assumption $s_1 > s_2$ of a simple intersection. On the other hand, the four dimensional system is not controlable from a single input variable and hence queues will in general not be dissolved at the same time. As a consequence, negative queue lengths will formally arise in the last phase of control treatment. In order to avoid this, state inequality constraints (5.36) should not be ignored in the problem solution /129/.

The control method for a single intersection just described cannot be viewded as a general control method for road networks. Research with respect to application of automatic control concepts to road networks still seems to be of a preliminary character /130, 131/. Simplification of the general problem formulation leading to reasonaable results with moderate implementation cost are still lacking and provide a future area of research. Nevertheless some pragmatic control systems for road networks based on heuristic considerations rather than on theoretical methods have been recently developed and successfully applied /132, 133/.

5.3 Control of freeway corridors

A long freeway is obviously only a part of a greater traffic system consisting of several freeways and surface streets. Optimal allocation of arriving traffic through the corridor system so as to meet an appropriate objective is a difficult large-scale problem. Open-loop and closed-loop, static and dynamic problem formulations have been proposed in the past by various researchers /83-88/. Since 1. the principal problems having to do with freeway corridors are similar to those of single freeways or street networks and 2. no general agreement about the possible solution approach has yet been achieved, we will not provi de a review of corridor control systems. The interested reader is referenced to the literature cited above.

6. THE MULTILEVEL APPROACH TO THE SOLUTION OF OPTIMAL CONTROL PROBLEMS.

In this chapter, some recently developed hierarchical algorithms for the solution of optimal control problems will be presented. Besides, results obtained by application of these algorithms to traffic control problems will be reported. Finally, implementation of a hierarchical optimization algorithm on a multi-microprocessor system will be described.

6.1 General notions

a) Introduction

Various hierarchical, multilevel methods for the solution of optimal control problems have been proposed over the last decade /89-93/. The most comprehensive treatment of the mathematical theory of hierarchical systems can be found in /89/. Application of a hierarchical method implies decomposition of the process under control into several interacting subprocesses, which provide a basis for the formulation of independent optimal control subproblems linked together by a separate coordination unit. A simple example /91/ will clarify these ideas.

Example 6.1

Suppose we wish to choose x and y to minimize the cost function

$$F = x^2 + 2xy + 2y^2 + 4 . \qquad (6.1)$$

If we use elementary calculus, we would set the derivatives of F with respect to (wrt) x and y to zero and solve the resulting simultaneous equations to get the obvious answer $x = y = 0$, $F = 4$.

Let us now structure this problem as a hierarchical problem to demonstrate the previous ideas. Define

$$F_1 = x^2 + x\,\pi_1 + 2 \tag{6.2}$$

$$F_2 = 2y^2 + \pi_2 y + 2 \tag{6.3}$$

where π_1 and π_2 represent the coupling between the two subsystems. The optimum values of x and y must then satisfy the interconnection constraints so that

$$\pi_1 = y \tag{6.4}$$

$$\pi_2 = x \tag{6.5}$$

and

$$F = F_1 + F_2. \tag{6.6}$$

Note that in general, $F \neq F_1 + F_2$ for values of x and y which do not satisfy the interconnection constraints. Now we have two infimal units S_1 and S_2 and two associated subproblems

S_1 : minimize F_1 wrt x

S_2 : minimize F_2 wrt y.

Before S_i, i = 1,2, can minimize F_i, however, a value must be supplied for π_i. This is the task of a supremal coordination unit S. The resulting structure is shown in figure 6.1. A possible solution procedure is the following :

Step 1: Set an initial guess for π_1, π_2, e.g. $\pi_1^1 = 2$, $\pi_2^1 = -3$. Set the iteration index L=1

Step 2: Solve independently problems S_1, S_2 and specify the solution x^L, y^L.

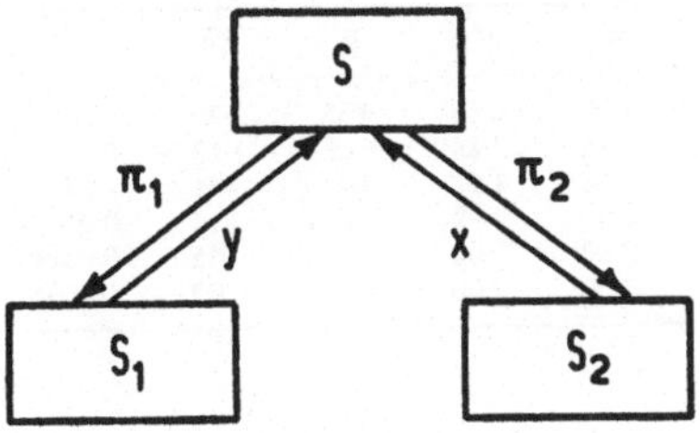

Figure 6.1 - Hierarchical optimization structure for example 6.1.

Step 3 : If an appropriately designed condition, e.g.

$$G = |\pi_1^L - y^L| + |\pi_2^L - x^L| < \varepsilon, \quad \varepsilon > 0 \tag{6.7}$$

is fulfilled, the overall problem is considered solved. If not, update π_1, π_2, for example through the relations

$$\pi_1^{L+1} = y^L \tag{6.8}$$

$$\pi_2^{L+1} = x^L \tag{6.9}$$

set L:=L+1 and go back to step 2.

The results of several iterations are shown in Table 6.1.x,y and F seem to be converging to 0,0 and 4 - the values we know to be optimum for this case[1].

This example demonstrates several interesting points. First, minimizing F_1 wrt x and F_2 wrt y is certainly easier than minimizing F wrt x and y. For example, in case of a large scale system, a system of nonlinear equations of high order should be solved simultaneously without decomposition.

---

(1) A convergence analysis can be easily performed for this simple example.

Iteration	π_1	π_2	x	y	F_1	F_2	F	G	ΔF
1	2·00	−3·00	−1·00	0·75	1·00	0·88	4·36	3·25	—
2	0·75	−1·00	−0·38	0·25	1·86	1·88	4·08	1·13	0·55
3	0·25	−0·38	−0·13	0·09	1·98	1·98	4·01	0·41	0·07
4	0·09	−0·13	−0·05	0·03	1·99	1·99	4·001	0·14	0·01
5	0·03	−0·05	−0·02	0·01	2·00	2·00	4·00	0·05	0·0011
6	0·01	−0·02	−0·006	0·004	2·00	2·00	4·00	0·02	0·0001

Table 6.1 -Iterative solution of example 1.1

Second, to obtain the overall optimal solution, we were required to iterate in some manner until the interconnection constraints were satisfied, i.e. until $\pi_1 = y$ and $\pi_2 = x$. Only if the interconnection constraints are satisfied are the two subproblems the same as the original problem. Third, it is important to note that the values of x and y which satisfy the interconnection constraints do not result in minimum values for F_1 and F_2. Finally, it should be noted that there is nothing in the procedure to guarantee convergence to the values of x and y which minimize F.

Of course, decomposition of the simple example 6.1 seems rather to complicate than to simplify the solution procedure. By application of hierarchical optimization techniques to large scale problems, however, the following main benefits are expected /94/:

(i) A substantial computation time reduction could be achieved for optimal control problems of sufficiently high order, especially if a multiprocessor system for the parallel execution of the independent subproblems is used. This feature might be extremely important for an on-line treatment of the optimization problem and hence for application of the results of optimal control theory to practical control problems.

(ii) Decomposition of the overall control problem into independent subproblems increases the number of variables involved in the solution procedure. As a consequence, more storage space in total is needed for the solution of the optimal control problem. However, in the case of decomposition, the total storage space can be distributed on several much smaller computers constituting a multicomputer system. On-line reassignment of the computers' tasks by a master computer has been proposed by some researchers /95/ leading to better reliability of the control system.

(iii) The decomposed problem consists of several separated modules which can be individually implemented and tested in a much more convenient way than in the case of a central treatment of the overall problem.

The question, if and up to which degree above benefits can be achieved when a hierarchical algorithm is applied to a specific optimal control problem, cannot be answered in general. The main factors affecting the efficiency of the control algorithm are on one hand the problem's order and structure and on the other hand the properties of the multicomputer system used. Hence, the development of hierarchical optimization algorithms has been mainly forced by

- the recent developments in microcomputer technology and
- the increasing of industrial complexes under control.

b) The general problem formulation

Consider the following discrete-time dynamical optimization overall problem

Minimize

$$J = \sum_{k=0}^{K-1} \phi \, [\underline{x}(k), \underline{u}(k), k] \qquad (6.10)$$

subject to the constraints

$$\underline{x}(k+1) = \underline{f} \, [\underline{x}(k), \underline{u}(k), k] \ ; \quad \underline{x}(o) = \underline{x}_o \qquad (6.11)$$

$$\underline{h} \, [\underline{x}(k), \underline{u}(k), k] \geq \underline{0} \qquad (6.12)$$

$$k = 0, \ldots, K-1; \quad K \text{ fixed},$$

where $x \in R^n$, $u \in R^m$ and inequality constraints (6.12) are assumed to satisfy the qualification condition given in /81/.

The optimal solution of the problem described by eqns. (6.10)-(6.12) must satisfy the necessary conditions given in Appendix A.2.2.b). The necessary conditions of optimality constitute a Two-Point-Boundary-Value-Problem(TPBVP), solution of which by use of iterative algorithms /42/ is an extremely difficult task, especially for large-scale processes. For this reason, we will try to decompose the overall problem into a number of independent subproblems.

c) The decomposed problem formulation

Let us assume that the system under control is decomposable into N interconnected subsystems as indicated in figure 6.2. The interconnection variables are generally given by

$$\underline{\pi}_i(k) = \sum_{j\neq i}^{N} \underline{g}_{ij} \, [\underline{x}_j(k), \, \underline{u}_j(k), \, k] \quad , \tag{6.13}$$

where $\underline{x}_j$, $\underline{u}_j$ are local state and control variables such that

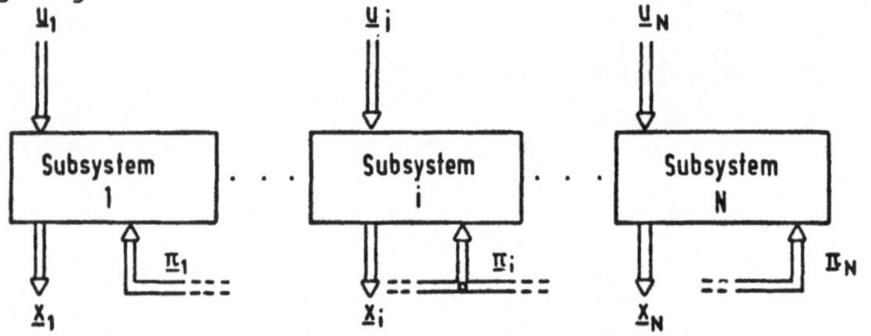

Figure 6.2 - A decomposed process

$$\underline{x} = [\underline{x}_1^T \; \dots \; \underline{x}_N^T]^T \; , \quad \underline{u} = [\underline{u}_1^T \dots \; \underline{u}_N^T]^T. \tag{6.14}$$

with above definitions, the overall problem given by eqns. (6.10)-(6.12) can be formulated in an equivalent decomposed form as follows:

Minimize

$$J = \sum_{k=0}^{K-1} \sum_{i=1}^{N} \phi_i \, [\underline{x}_i(k),\ \underline{u}_i(k),\ \underline{\pi}_i(k),\ k] \qquad (6.15)$$

subject to

$$\underline{x}_i(k+1) = \underline{f}_i \, [\underline{x}_i(k),\ \underline{u}_i(k),\ \underline{\pi}_i(k),\ k]\ ,\ \underline{x}_i(0) = \underline{x}_{io} \qquad (6.16)$$

$$\underline{h}_i \, [\underline{x}_i(k),\ \underline{u}_i(k),\ \underline{\pi}_i(k),\ k\,] \geq \underline{0} \qquad (6.17)$$

and eqn. (6.13)

$k = 0,\ldots, K-1$, K fixed; $i = 1,\ldots, N$.

The hamiltonian of the overall problem in decomposed form is formulated by adjoining the interconnection constraints (6.13) with an additional Lagrange multiplier vector $\underline{\beta}_i$ to the usual Hamiltonian, i.e.

$$H = \sum_{i=1}^{N} \{\phi_i \, [\underline{x}_i(k),\ \underline{u}_i(k),\ \underline{\pi}_i(k),\ k] + \underline{\lambda}_i(k+1)^T \, \underline{f}_i \, [\underline{x}_i(k),\ \underline{u}_i(k),$$
$$\underline{\pi}_i(k),\ k] + \underline{\beta}_i(k)^T \, [\underline{\pi}_i(k) - \sum_{j\neq i}^{N} \underline{g}_{ij}[\,\underline{x}_j(k), \underline{u}_j(k),\ k]]\} \qquad (6.18)$$

The basic idea of hierarchical optimization theory is to decompose the overall Hamiltonian, given by equation (6.18), and formulate several subproblems of lower dimension than the original problem given. Then the subproblems could be solved independently subject to some global variables provided by a coordination unit. In order to accomplish this, only local state and control variables $\underline{x}_i(k)$ and $\underline{u}_i(k)$ must appear in the equations of the subproblems. This can be achieved in our case if the third term of the right hand side of eqn. (6.18) is rearranged so that:

$$H = \sum_{i=1}^{N} \{\phi_i \, [\underline{x}_i(k),\ \underline{u}_i(k),\ \underline{\pi}_i(k),\ k\,] + \underline{\lambda}_i(k+1)^T \underline{f}_i \, [\underline{x}_i(k), \underline{u}_i(k),$$
$$\underline{\pi}_i(k),\ \ k)] + \underline{\beta}_i(k)^T \, \underline{\pi}_i(k) - \sum_{j\neq i}^{N} \underline{\beta}_j(k)^T \underline{g}_{ji} \, [\underline{x}_i(k), \underline{u}_i(k), k]\},$$

(6.19)

which means that

$$H = \sum_{i=1}^{N} H_i \; [\underline{x}_i(k),\ \underline{u}_i(k),\ \underline{\pi}_i(k),\ \underline{\beta}(k),\ k] \;, \qquad (6.20)$$

i.e. only $\underline{\beta}$ appears as a global variable of the sub-Hamiltonians H_i.

Necessary conditions for optimality of the overall problem can be derived by use of the Hamiltonian (6.19)

$$\underline{x}_i(k+1) = \frac{\partial H_i}{\partial \underline{\lambda}_i} = \underline{f}_i \; [\underline{x}_i(k),\ \underline{u}_i(k),\ \underline{\pi}_i(k),\ k] \;;\quad \underline{x}_i(0)=\underline{x}_{io} \qquad (6.21)$$

$$\underline{\lambda}_i(k) = \frac{\partial H_i}{\partial \underline{x}_i(k)} + \frac{\partial \underline{h}_i^T}{\partial \underline{x}_i(k)} \underline{\mu}_i(k) = \frac{\partial \phi_i}{\partial \underline{x}_i(k)} + \frac{\partial \underline{f}_i^T}{\partial \underline{x}_i(k)} \underline{\lambda}_i(k+1) -$$

$$- \sum_{j \neq i}^{N} \frac{\partial \underline{g}_{ji}^T}{\partial \underline{x}_i(k)} \underline{\beta}_j(k) + \frac{\partial \underline{h}_i^T}{\partial \underline{x}_i(k)} \underline{\mu}_i(k);$$

$$\underline{\lambda}_i(K) = \underline{0} \qquad (6.22)$$

$$\frac{\partial H}{\partial \underline{u}_i(k)} + \frac{\partial \underline{h}_i^T}{\partial \underline{u}_i(k)} \underline{\mu}_i(k) = \frac{\partial \phi_i}{\partial \underline{u}_i(k)} + \frac{\partial \underline{f}_i^T}{\partial \underline{u}_i(k)} \underline{\lambda}_i(k+1) -$$

$$\sum_{j \neq i}^{N} \frac{\partial \underline{g}_{ji}^T}{\partial \underline{u}_i(k)} \underline{\beta}_j(k) + \frac{\partial \underline{h}_i^T}{\partial \underline{u}_i(k)} \underline{\mu}_i(k) = \underline{0} \qquad (6.23)$$

$$\underline{\mu}_i(k)^T \underline{h}_i(k) = 0 \;;\quad \underline{\mu}_i(k) \leq \underline{0} \;;\quad \underline{h}_i \; [\underline{x}_i(k),\ \underline{u}_i(k),\ \underline{\pi}_i(k), k] \geq \underline{0}$$

(6.24)

$$\frac{\partial H}{\partial \underline{\beta}_i(k)} = 0 \iff \underline{\pi}_i(k) = \sum_{j \neq i}^{N} \underline{g}_{ij} \; [\underline{x}_j(k),\ \underline{u}_j(k), k] \qquad (6.25)$$

$$\frac{\partial H}{\partial \underline{\pi}_i(k)} + \frac{\partial \underline{h}_i^T}{\partial \underline{\pi}_i(k)} \underline{\mu}_i(k) = \underline{0} \qquad \iff$$

$$\underline{\beta}_i(k) = -\frac{\partial \phi_i}{\partial \underline{\pi}_i(k)} - \frac{\partial \underline{f}_i^T}{\partial \underline{\pi}_i(k)} \underline{\lambda}_i(k+1) - \frac{\partial \underline{h}_i^T}{\partial \underline{\pi}_i(k)} \underline{\mu}_i(k) \qquad (6.26)$$

$k = 0, \ldots, K-1; \quad i = 1, \ldots, N.$

d) The interaction prediction principle

Assume that we can in some way predict the values of the trajectories $\underline{\pi}(k)$ and $\underline{\beta}(k)$ which will be called the coordination variables. If $\underline{\pi}(k)$ and $\underline{\beta}(k)$ are treated as known inputs, it is easily seen that the overall problem is composed of N independent subproblems, the set up of which can be deduced by the sub-Hamiltonians of eqn. (6.19):

$$\underset{\underline{x}_i,\, \underline{u}_i}{\text{Minimize}} \quad J_i = \sum_{k=0}^{K-1} \{\phi_i[\underline{x}_i(k), \underline{u}_i(k), \underline{\pi}_i(k), k] + \underline{\beta}_i(k)^T \underline{\pi}_i(k) - \sum_{j \neq i}^{N} \underline{\beta}_j(k)^T \underline{g}_{ji}[\underline{x}_i(k), \underline{u}_i(k)\; k]\} \qquad (6.27)$$

subject to eqns. (6.16), (6.17).

The necessary conditions for optimality of these independent suproblems are given by eqns. (6.21) - (6.24). Hence, if we can solve above subproblems for some values of $\underline{\pi}(k)$ and $\underline{\beta}(k)$ satisfying eqns. (6.25), (6.26), the overall problem will be solved.

It is thus apparent that an important problem in hierarchical system theory is the determination of the coordination variables $\underline{\pi}(k)$ and $\underline{\beta}(k)$ by a suppremal coordination unit. Since specification of $\underline{\pi}(k)$ and $\underline{\beta}(k)$ by use of eqns. (6.25), (6.26) requires knowledge of the optimal local variables and vice versa, an iterative procedure must be followed:

Step 1 : Guess some trajectories for the coordination variables $\underline{\pi}^1(k)$, $\underline{\beta}^1(k)$. Set the iteration index $L = 1$.

Step 2 : Solve N independent subproblems for given coordination variables and specify the solution trajectories of the local variables $\underline{x}_i^L$, $\underline{u}_i^L$, $\underline{\lambda}_i^L$, $\underline{\mu}_i^L$.

Step 3 : Update the coordination variables by use of the last subproblems' solutions.

Step 4 : If

$$||\underline{\pi}^{L+1} - \underline{\pi}^L|| + ||\underline{\beta}^{L+1} - \underline{\beta}^L|| > \varepsilon \tag{6.28}$$

for some prescribed accuracy variable $\varepsilon > 0$, set $L: = L+1$ and go to step 2, else stop and record the actual local variables as the optimal solution of the overall control problem.

The solution structure is shown in figure 6.3. Updating of the coordination variables at step 3 can be performed by direct use of eqns. (6.25), (6.26). In order to increase the region of attraction for convergence, some smoothing might be useful

$$\underline{\pi}^{L+1}(k) \;:=\; \sigma .\ \underline{\pi}^{L+1}(k) + (1-\sigma)\ \underline{\pi}^L(k) \tag{6.29}$$

$$\underline{\beta}^{L+1}(k) \;:=\; \sigma .\ \underline{\beta}^{L+1}(k) + (1-\sigma)\ \underline{\beta}^L(k), \tag{6.30}$$

$0 < \sigma \leq 1.$

Alternatively, gradient techniques for minimization resp. maximization of H can be used

$$\underline{\pi}^{L+1}(k) = \underline{\pi}^L(k) - K_\pi^L \left[\frac{\partial H^L}{\partial \underline{\pi}} + \frac{(\partial \underline{h}^L)^T}{\partial \underline{\pi}}\ \underline{\mu}^L(k)\right] \tag{6.31}$$

$$\underline{\beta}^{L+1}(k) = \underline{\beta}^L(k) + K_\beta^L \cdot \frac{\partial H^L}{\partial \underline{\beta}} \tag{6.32}$$

K_π^L , $K_\beta^L > 0.$

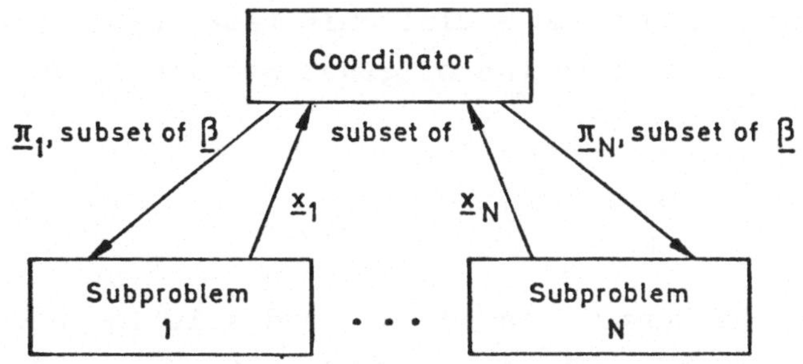

Figure 6.3 - Decomposed optimization structure.

The signs in eqns. (6.31), (6.32) have been set according to the relationships between necessary conditions and the saddle point conditions of the corresponding Min Max problem /96/. Convergence proofs of above iterative algorithm can be given only for particular cases.

e) The interaction balance principle

According to eqn. (6.20), $\underline{\beta}(k)$ is the only global variable included in the sub-Hamiltonians H_i. Hence, an alternative hierarchical algorithm can be constructed by treating the interaction variables $\underline{\pi}_i(k)$ as additional input variables for the subproblems. In that case, minimization of (6.27) should be performed with respect to $\underline{x}_i$, $\underline{u}_i$ and $\underline{\pi}_i(k)$ so that the necessary conditions of the subproblems are given by eqns. (6.21) - (6.24), (6.26). Then, the only equation which must be fulfilled by the coordinator is equation (6.25) and this can be done by use of the gradient update, eqn. (6.32). The overall problem is solved in a similar iterative way as for the interaction prediction principle.

Difficulties may arise by application of the balance principle, if the sub-Hamiltonians are linear in $\underline{\pi}_i$. In that case, solution of the subproblems may become a difficult task, even if no such difficulties were present in the original problem formulation.

f) Summary

Let us summarize the central and decomposed solution procedure. In the central case, a large-scale TPBVP must be solved in order to specify the optimal solution. Since this is a nontrivial task, we decompose the process under control into interconnected subprocesses and formulate the overall optimization problem in decomposed form. By dividing the problem variables into local variables and coordination variables, we are able to decompose the overall problem into a number of independent subproblems associated with the individual subprocesses. Solution of the resulting low dimensional TPBVPs in an iterational procedure leads to the overall solution. Possible computation time and storage space savings obtained by this hierarchical treatment are discussed in section 6.3.

6.2 Application to traffic control

a) Freeway traffic control

We will first apply the presented algorithms to the freeway traffic control problem. We will consider an optimization problem similar to P3 of section 5.1e). The main modification compared to P3 will be the consideration of a quadratic performance functional

$$J = \frac{1}{2} \sum_{k=0}^{K-1} \{ \|\underline{x}(k) - \underline{x}_N(k)\|_Q^2 + \|\underline{u}(k) - \underline{u}_N(k)\|_R^2 \} \tag{6.33}$$

where $Q, R \geq 0$ and $\underline{x}_N(k)$, $\underline{u}_N(k)$ are some desired state and input variable trajectories (or nominal points) which are supposed to be known. For example, $\underline{x}_N$, $\underline{u}_N$ can be derived by solution of a steady-state (section 5.1c) or dynamic (section 5.1d) time-of-day control problem (see also chapter 7 for details). Introduction of the quadratic performance criterion (6.33) is necessary because numerical solution of problem P3 eather in a central or in a decomposed manner has been found to be an extremely difficult task. The reasons for that have been outlined in section 5.1.e).

The second modification of problem P3 is that we consider only on-ramp volumes r as input variables of the system, whilst speed limitation b are set equal to their nominal values. Consideration of b in the performance criterion (6.33) should lead to a rise of the total travel time in some situations /67/. Finally, for the sake of simplicity, no fixed upper limits for the on-ramp volumes and the queue lengths are consider, i.e. inequality constraints (5.7) become

$$0 \leqslant \underline{r}_{min} \leqslant \underline{r}(k) \leqslant \underline{d}(k) + \frac{1}{T}\,\underline{l}(k). \qquad (6.34)$$

Furthermore, we will assume that the freeway under consideration is divided into N sections each including at most one on-ramp and one off-ramp. Each section is furthermore subdivided into n(i), i = 1, ..., N, segments. The traffic density and mean speed in the j-th segment of the i-th section are denoted by ρ_i^j, v_i^j. The state vector is given by

$$\underline{x}^T = [l_1 \; \rho_1^1 \; v_1^1 \; \dots \; \rho_N^{n(N)} \; v_N^{n(N)}] \qquad (6.35)$$

and the input vector by

$$\underline{u}^T = [r_1 \dots r_N] \; . \qquad (6.36)$$

Off-ramp volumes are considered through eqn. (2.16) assuming known off-ramp rates $\gamma(k)$. Finally the weighting factor α of eqn. (3.3) is set to one.

Find $\underline{u}^*(k)$, $k = 0, \ldots, K-1$,
so as to minimize J given by eqn. (6.33)
subject to eqns. (3.1) - (3.3) taken for each freeway segment and eqns. (5.2), (6.34) taken for each freeway section.

By inspection of the Hamiltonian of the problem, it can be easily seen that introduction of the convex cost functional (6.33) leads to a unique choice of the optimal input variables. In fact, the terms of the Hamiltonian depending upon on-ramp volume $r_i(k)$ are

$$H = \ldots + \frac{r}{2}\left[r_i(k) - r_{i,N}(k)\right]^2 - T\, r_i(k)\, \lambda_i^1(k+1) + \frac{T}{\Delta_i^1} r_i(k)\, \lambda_i^2(k+1) + \ldots \tag{6.37}$$

where $R = r.E$ has been set. On the other hand, inequality constraints (6.34) have the form of (A2.15) and hence the optimal value of the on-ramp volumes is selected according to the sign of $\frac{\partial H}{\partial r_i}$ at the minimum as required by eqn. (A2.16). It is straightforward to show that the optimal control rule in our case is given by

$$r_i^*(k) = \begin{cases} r_{i,min} & \text{if } r_i^m(k) \leq r_{i,min} \\ r_i^m(k) & \text{if } r_{i,min} < r_i^m(k) < d_i(k) + \frac{1}{T} l_i^*(k) \\ d_i(k) + \frac{1}{T} l_i^*(k) & \text{if } r_i^m(k) \geq d_i(k) + \frac{1}{T} l_i^*(k) \end{cases} \tag{6.38}$$

where

$$r_i^m(k) = r_{i,N}(k) + \frac{T}{r}\, \lambda_i^1(k+1) - \frac{T}{r.\, \Delta_i^1}\, \lambda_i^2(k+1) \tag{6.39}$$

has been derived from the condition $\frac{\partial H}{\partial r_i} = 0$.

The order of the state vector is 2n+N and grows proportionally to the length of the freeway under consideration. The order of the associated TPBVP raises accordingly. The overall problem can be subdivided into N independent subproblems corresponding to particular freeway section by introducing the interconnection variables /97, 98/ (figure 6.4)

$$\underline{\pi}_i^T = [\, \rho_{i-1}^{n(i-1)} \quad v_{i-1}^{n(i-1)} \quad \rho_{i+1}^{1} \,] \tag{6.40}$$

and the local state and control variables

$$\underline{x}_i^T = [l_i \; \rho_i^1 \; v_i^1 \; \dots \; \rho_i^{n(i)} v_i^{n(i)}] \tag{6.41}$$

$$u_i = r_i \; . \tag{6.42}$$

On the base of this process decomposition, the multilevel algorithms described in section 6.1 can be applied.

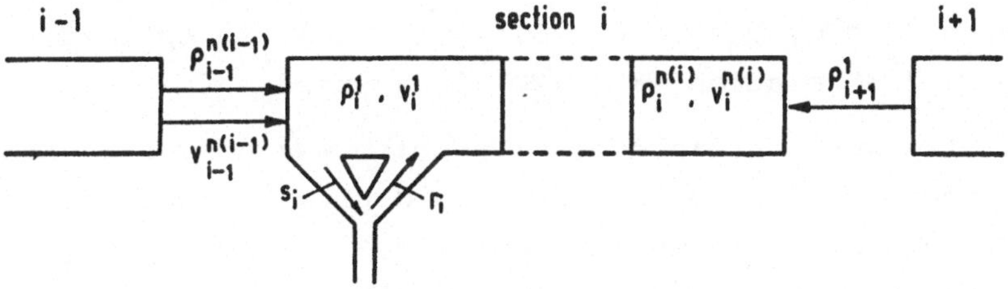

Figure 6.4 Interconnections $\underline{\pi}_i$ for a freeway section.

In order to study the efficiency of the central and decomposed algorithms and the properties of the optimal traffic control strategies, a traffic situation as it often occurs in daily traffic has been simulated on a digital computer. A hypothetical six-section, two-lane freeway with six on-ramps and six off-ramps has been considered. Each section is 5 km long and is subdivided into five segments of 1 km in legnth.
It is assumed that congestion is present at time zero in the 2nd segment of the 3rd section caused by an accident or some other severe disturbance of traffic flow. We consider a desi-

red operation point given by

$$\rho^j_{1,N} = 37 \qquad \rho^j_{2,N} = 49 \qquad \rho^j_{i,N} = 64 \;, \qquad i = 3,.., 6 \;\; ; \; j = 1,...,5$$

$$v^j_{1,N} = 82 \qquad v^j_{2,N} = 74 \qquad v^j_{i,N} = 62 \;, \qquad i = 3,...,6 \;\; ; \; j = 1,...,5$$

$$r_{1,N} = 3000, \; r_{2,N} = 700, \; r_{3,N} = 635 \;, \; r_{4,N} = 217 \;, \; r_{5,N} = 228 \;, \; r_{6,N} = 239$$

and the corresponding demands

$$d_{1,N} = 3000; \quad d_{2,N} = d_{3,N} = 700; \quad d_{4,N} = d_{5,N} = d_{6,N} = 350$$

remaining constant over the whole period of control operation. The exit rates are set to

$$\gamma_{1,N} = 0, \quad \gamma_{2,N} = 0{,}05, \; \gamma_{3,N} = 0.052 \;; \; \gamma_{4,N} = 0.054 \;; \; \gamma_{5,N} = 0{,}057$$

$$\gamma_{6,N} = 0.059$$

and the weighting matrices $Q = \text{diag}\ (Q_1, Q_\rho, Q_v, \ldots, Q_\rho, Q_v)$ $R = v.E$ with $Q_1 = 0$, $Q_\rho = 1$, $Q_v = 0{,}3$, $r = 0.01$. An optimization time horizon of 30 minutes ($K = 120$) is assumed.

The optimal freeway traffic control problem was solved in its central and decomposed form on a CYBER 175 digital computer of the Leibnitz Rechenzentrum in Munich. Results are shown in figure 6.5. Application of nonlinear optimal control drives traffic back to its nominal condition after occurence of a severe disturbance. The initial overcritical traffic density has been decreased to values near the maximum flow point.

The state vector of the optimal traffic control problem given in eqn. (6.41) is of 66th order and the control vector is of 6th order. As described above, the overall problem can be decomposed into six subproblems with 11-dimensional local state vectors, scalar local control variables and 3-dimensional interaction vectors. Results obtained by implementation of the

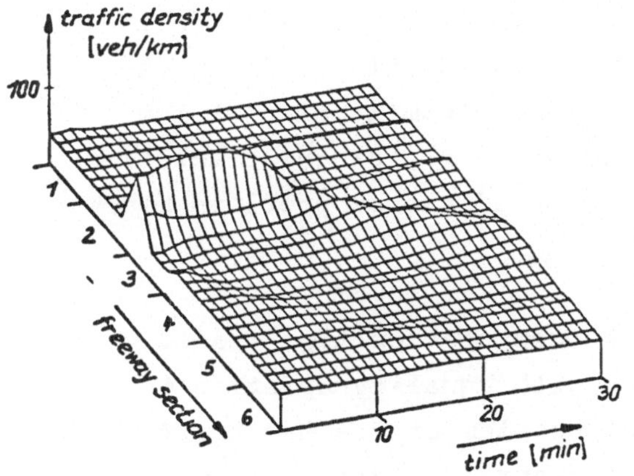

Figure 6.5 -
Evolution of traffic density for optimal non-linear control.

decomposed problem on a multi-microprocessor-system will be reported in section 6.3. We will here present the results obtained on a single CYBER 175 digital computer by application of several hierarcical methods.

Table 6.1 reviews the results obtained. In particular :
(i) Prediction principle with equality updating in the coordinator provides the best results for the decomposed case. Implementation of the interaction prediction algorithm on a multi-processor-system is expected to reduce the total computation time as compared to the central solution. The numbers of second-level iterations (=16) is rather low.

(ii) Prediction principle with gradient updating in the coordinator creates difficulties, since it seems to be extremely difficult to determine a single gradient step avoiding divergence and leading to quick convergence of the iterative procedure. Extremely long computation time is the consequence.

Method	Central solution	Interaction prediction eq .update	gradient update	Interaction balance
Computat-tion time /sec/	20	58	too slow	singular subproblems

Table 6.1 - Computation time for various solution methods.

(iii) By use of the interaction balance principle, the interaction variables $\underline{\pi}_i$ are treated as additional input variables of the subproblems. Since the hamiltonian of the subproblems, eqn. (6.19) is linear in $\underline{\pi}_i$, singular solutions result. As a consequence, solution of the subproblems becomes a very difficult task.

Above results indicate that the computation time needed for the prediction principle algorithm with equality updating to solve the nonlinear optimal control problem of freeway traffic is comparable with the one of the central solution. In section 6.3 some more details concerning implementation of this algorithm on a multiprocessor system will be given. On the other hand, interaction prediction with gradient updating, interaction balance and further multilevel algorithms investigated in /97, 98/ fail to provide satisfactory results.

b) Road traffic Control

b1) The optimal control problem

Following the treatment of /99/ we will now consider the problem of optimal control of road networks. Eqns. (2.38), (2.39) provide the state equations of a simple intersection and eqns. (2.41) - (2.44) provide the state equations for a particular complex intersection. It is easily seen that the state equations of a single intersection have the general form

$$\dot{\underline{l}} = \bar{\underline{b}} \cdot r + \bar{\underline{a}} \qquad (6.43)$$

with $\underline{l}$ the state vector, the dimension of which depends upon the intersection's complexity, r a scalar input and $\bar{a}$ a time varying vector. The discrete time version of eqn. (6.43) is given by:

$$\underline{l}(k+1) = A \cdot \underline{l}(k) + \underline{b} \cdot r(k) + \underline{a}(k) \qquad (6.44)$$

where A is equal to the identity matrix E. As a sampling period, it is reasonable to use a cycle time period t_c.

Consider the case of two subsequent intersections shown in figure 6.6. It is obvious that arriving rate $d_3(k)$ in the second intersection depends somehow on the serving rate $r_1(k)$ (1) of direction 1 of the first intersection. Since, for the oversaturated junctions only the macrobehaviour is important, it is adquate when considering whole networks to model interconnecting roads as pure delay elements. Then, we can write for the particular case of figure 6.6

$$d_3(k) = r_1(k-m) \qquad (6.45)$$

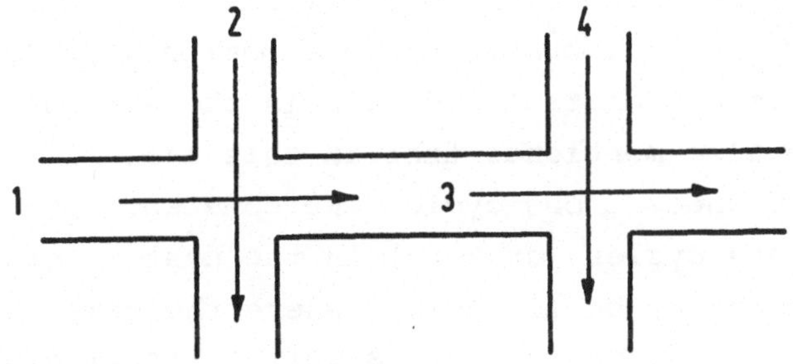

Figure 6.6 - A model for the interconnected road.

where m is the number of delay periods. It can be easily seen, that the overall network can be represented in that manner by a linear vector matrix difference equation with pure delays in the controls having the general form

$$\underline{l}(k+1) = A \cdot \underline{l}(k) + \sum_{\nu=o}^{m} B_{\nu} \cdot \underline{r}(k-\nu) + \underline{a}(k) \qquad (6.46)$$

(1) The index of r(k) denotes the number of the intersection.

where $\underline{l}$ is the vector of the queues on all the arms of all junctions, A is equal to the identity matrix, B_ν , $\nu = 0, \ldots, m$, are the input matrices, $\underline{r}$ is the vector of control inputs, the dimension of which equals the number of intersections considered and $\underline{a}$ is a time varying vector including the arriving rates coming from outside the network boundary.

At this point it should be mentioned that there are two types of control variables used for road networks in traffic engi neering practice. These are (i) the 'split' which effectively corresponds to the above controls r_i for each intersection and (ii) the 'offset' which is the time difference between the start of the green of one intersection relative to the start of green on another. There are in fact two possible ways of treating the offsets; they can be treated eather as constants or as additional control variables. Now since for oversaturated intersections only the macrobehaviour is important, it is realistic to treat them as constant and this is the approach used in/99/.It should be also mentioned that in this analysis, since the sampling period used is one cycle, the constant offsets are taken as complete cycles (or zero) in the system description. There may, however, be situations where the time delays between one intersection and another is only a fraction of the cycle time and this is then the constant offset. In that case also, the same analysis applies although a smaller sample period must then be used. In such situations, if desired, it is easy to include the offset as additional controls of the system.

For reasons given in section 5.2, the states and controls are subject to inequality constraints (5.31) - (5.33). A cost functional similar to the one of eqn. (5.35) has been considered in /99/

$$J = \frac{1}{2} \sum_{k=0}^{K-1} \{ \| \underline{l}(k) \|_Q^2 + \| \underline{r}(k) - \underline{r}_N \|_R^2 \}; \quad K \text{ fixed} \tag{6.47}$$

where the second term in the sum has been added to facilitate the solution procedure. $\underline{r}_N$ are some desired control inputs and diagonal weighting matrix R should be chosen "small" compared to the weighting matrix Q of the states.

Thus we are in a position to formulate the optimal control problem P6 for a road network :

Given $\underline{l}(0) = \underline{l}_0$, $\underline{r}(k) = \underline{0} \; \forall k < 0$
Minimize J, eqn. (6.47)
subject to eqn. (6.46)
and inequality constraints (5.31) - (5.33).

b2) The solution procedure

Traditionally, problems with time delays are treated by augmenting the state space introducing additional variables for the delay terms and converting the original high-order difference equation to a set of coupled first order equations.
Since this approach increases the dimensionality, a different treatment has been proposed in /99/. Although the solution procedure described below has been presented in the context of hierarchical optimization, it can not be understood as a multilevel algorithm in the sense of section 6.1.

Considering P6 as a static optimization problem [1)] we can write the Lagrangian

---

(1) Obviously, every discrete time optimization problem can be considered as a static one /81/.

$$L(\underline{l}, \underline{r}, \underline{\lambda}) = \sum_{K=0}^{K-1} \{ \tfrac{1}{2}[\| \underline{l}(k) \|_Q^2 + \| \underline{r}(k) - \underline{r}_N \|_R^2] + \underline{\lambda}(k+1)^T [\underline{l}(k+1) -$$

$$- A\, \underline{l}(k) - \sum_{\nu=0}^{m} B_\nu \cdot \underline{r}\,(k-\nu) - \underline{a}(k)]\} . \qquad (6.48)$$

Since J is a convex function and the constraints are linear, a necessary and sufficient condition for the solution of P6 is to solve the minmax problem /100/

$$\max_{\underline{\lambda}} \min_{\underline{l},\underline{r}} \{L(\underline{l},\underline{r},\underline{\lambda}) \text{ subject to eqns. } (5.31)-(5.33)\} . \qquad (6.49)$$

Solution of the minmax problem can be obtained by the following iterative procedure

Step 1 : Guess an initial trajectory of the Lagrange multipliers $\underline{\lambda}^1(k)$, $k = 1, \ldots, K$. Set the iteration index $L = 1$.

Step 2 : Minimize $L(\underline{l},\underline{r},\underline{\lambda}^L)$ subject to (5.31) - (5.33) and specify the solution $\underline{l}^L$, $\underline{r}^L$.

Step 3 : Calculate the gradient of $L(\underline{l}^L, \underline{r}^L, \underline{\lambda}^L)$ w.r.t. $\underline{\lambda}$ and improve the $\underline{\lambda}$-trajectory using say the conjugate gradient algorithm so that

$$\underline{\lambda}^{L+1} = \underline{\lambda}^L + \alpha^L \cdot \underline{d}^L \qquad (6.50)$$

where $\underline{d}^L$ is the conjugate direction and α^L is the step length. The search direction $\underline{d}^L$ can be calculated using the relationship

$$\underline{d}^L = \underline{\Delta}^L + \beta^{L-1}\, \underline{d}^{L-1} \qquad (6.51)$$

where

$$\beta^L = (\underline{\Delta}^L)^T\, \underline{\Delta}^L / (\underline{\Delta}^{L-1})^T\, \underline{\Delta}^{L-1} \qquad (6.52)$$

---

(1) Obviously, every discrete time optimization problem can be considered as a static one /81/.

and $\underline{\Delta}$ is the gradient of $L(\underline{l}, \underline{r}, \underline{\lambda})$ w.r.t. $\underline{\lambda}$. Initially, $\underline{d}^1 = \underline{\Delta}^1$, i.e. the steepest ascent direction.

Step 4 : If $\|\underline{\Delta}^L\| > \varepsilon$ for some prescribed accuracy variable ε, set L:=L+1 and go to step 2, else stop and record the actual solution of step 2 as the optimal one.

Solution of the minimization problem in step 2 is an easy task in spite of the presence of inequality constraints, since the minimization can be performed independently for each variable and each time instance. To see this, rewrite eqn. (6.48) in the form

$$L(\underline{l},\underline{r},\underline{\lambda}) = \sum_{k=0}^{K-1} \{\frac{1}{2} [\|\underline{l}(k)\|_Q^2 + \|\underline{r}(k)-\underline{r}_N\|_R^2] + \underline{\lambda}(k)^T \underline{l}(k) - \underline{\lambda}(k+1)^T [A\,\underline{l}(k)+\underline{a}(k)] - \sum_{\nu=0}^{m} \underline{\lambda}(k+1+\nu)^T B_\nu \underline{r}(k)\} + \underline{\lambda}(K)^T \underline{l}(K) \qquad (6.53)$$

where $\underline{\lambda}(K+\nu) = \underline{0}$, $\nu > 0$ and $\underline{\lambda}(0) = \underline{0}$.

Since $L(\underline{l},\underline{r}, \underline{\lambda})$ is convex, minimization results can be given analytically :

$$\underline{r}^L(k) = \text{sat}\,\{\underline{r}_N+R^{-1} \sum_{\nu=0}^{m} B_\nu^T \underline{\lambda}^L(k+1+\nu)\}\ , \ k = 0, \ldots, K-1, \qquad (6.54)$$

$$\underline{l}^L(k) = \text{sat}\,\{Q^{-1} [-\underline{\lambda}^L(k)+A^T \underline{\lambda}^L(k+1)]\}, \quad k=1,\ldots,K \qquad (6.55)$$

where

$$\text{sat}\{\eta\} = \begin{cases} \eta_{max} & \text{if} \quad \eta \geq \eta_{max} \\ \eta & \text{if} \quad \eta_{min} < \eta < \eta_{max} \\ \eta_{min} & \text{if} \quad \eta \leq \eta_{min} \end{cases} \qquad (6.56)$$

and saturation of a vector $\underline{\eta}$ is the vector of the saturations of the elements of $\underline{\eta}$.

The gradient $\underline{\Delta}$ of the Lagrangian w.r.t. $\underline{\lambda}$ needed in step 3 is easily derived from eqn. (6.48)

$$\underline{\Delta}^L(k+1) = \underline{1}^L(K+1) - A\,\underline{1}^L(k) - \sum_{\nu=0}^{m} B_\nu\,\underline{r}^L(k-\nu) - \underline{a}(k). \quad (6.57)$$

Thus the gradient $\underline{\Delta}$ is simply the error in the system equation during the iterative procedure.

b3) Numerical study

As an example consider a small London network shown in figure 6.7. The network consists of three intersections and is a major trouble spot in the West London area. The difficulties arise because the intersections are very close, both to each other and to the neighbouring intersections, so that the storage available

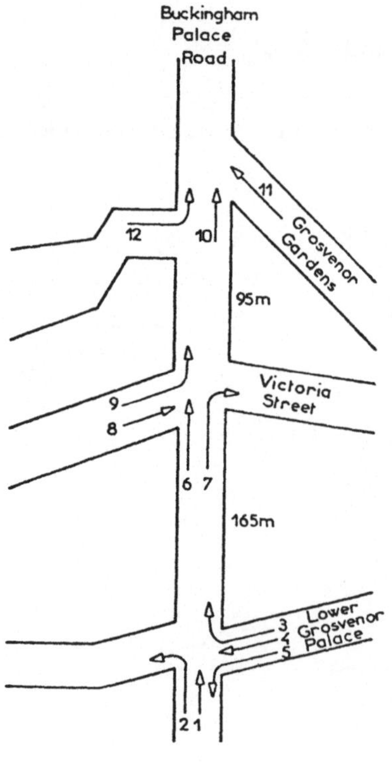

Fig. 6.7 -
A London network.

on the linking roads is fairly small. The state inequality constraints cannot therefore be relaxed. Based on the existing control structure, the system has the following three controls:

r_1 serving rate associated with a green phase for streams 1 and 2

r_2 serving rate associated with a green phase for streams 6 and 7 and with a two-cycle delay with respect to r_1

r_3 serving rate associated with a green phase for streams 10 and 12 and with an one-cycle delay with respect to r_2.

The cycle time for all intersections is 1 min and the loss time is 6 sec. Considering $u_i = g_i/t_c = s_i \cdot r_i$ as new input variables (see eqn. (2.34)) the linear structure of eqn. (6.46) is being preserved. u_i are the cycle fractions taken as green phases for the mentioned directions.

Desired values for the controls were chosen to be $u_{1,N} = 0.45$, $u_{2,N} = 0.3$, $u_{3,N} = 0.5$. We define

$$\Delta u_i = u_i - u_{i,N}. \tag{6.58}$$

Using data of inflows into this network and saturation flows as measured between 17.00 and 18.00 h by the Greater London Council /99/, the state euqations can be written in vector matrix form as

$$\underline{l}(k+1) = E \cdot \underline{l}(k) + B_o \cdot \Delta\underline{u}(k) + B_1 \cdot \Delta\underline{u}(k-1) + B_2 \Delta\underline{u}(k-2) + \underline{a} \tag{6.59}$$

where E is the twelfth-order identity matrix and

$$B_o = \begin{bmatrix} -65 & 0 & 0 \\ -25 & 0 & 0 \\ 34 & 0 & 0 \\ 31 & 0 & 0 \\ 4 & 0 & 0 \\ 0 & -64 & 0 \\ 0 & -26 & 0 \\ 0 & 132 & 0 \\ 0 & 34 & 0 \\ 0 & 0 & -96 \\ 0 & 0 & 90 \\ 0 & 0 & -25 \end{bmatrix}, \quad B_1 = \begin{bmatrix} 0 & 0 & 0 \\ 0 & 0 & 0 \\ 0 & 0 & 0 \\ 0 & 0 & 0 \\ 0 & 0 & 0 \\ 0 & 0 & 0 \\ 0 & 0 & 0 \\ 0 & 0 & 0 \\ 0 & 0 & 0 \\ 0 & 30 & 0 \\ 0 & 0 & 0 \\ 0 & 0 & 0 \end{bmatrix}, \quad B_2 = \begin{bmatrix} 0 & 0 & 0 \\ 0 & 0 & 0 \\ 0 & 0 & 0 \\ 0 & 0 & 0 \\ 42.7 & 0 & 0 \\ 18.3 & 0 & 0 \\ 0 & 0 & 0 \\ 0 & 0 & 0 \\ 0 & 0 & 0 \\ 0 & 0 & 0 \\ 0 & 0 & 0 \\ 0 & 0 & 0 \end{bmatrix}$$

$$\underline{a} = [-21.6 \quad -8.2 \quad 8.4 \quad 7.7 \quad 0.9 \quad 2.56 \quad 1.62 \quad -64.2 \quad -16.4 \quad -8.4 \quad -33.6 \quad -10.5\,]^T.$$

The cost function was chosen to be

$$J = \sum_{k=0}^{2} \left[\|\underline{l}(k)\|_Q^2 + 100\,\|\Delta\underline{u}(k)\|_R^2 \right]$$

where $R = E$ and $Q = \mathrm{diag}\{1, 1, 1, 1, 1, 1.5, 1.5, 1.1, 2, 1, 2\}$.

The states l_6, l_7, l_{10} are favoured because of the limited storage on the interconnection roads between the intersections. The state and control variable constraints were chosen to be

$$\begin{aligned} &0 \leq l_i \leq 40\,, && i = 1, 2, 8, 9, 11, 12 \\ &0 \leq l_i \leq 80\,, && i = 3, 4, 5 \\ &0 \leq l_i \leq 50\,, && i = 6,7 \\ &0 \leq l_{10} \leq 25 && \end{aligned}$$

$-0.25 \leqslant \Delta u_1 \leqslant 0.25$, $-0.1 \leqslant \Delta u_2 \leqslant 0.4$, $-0.3 \leqslant \Delta u_3 \leqslant 0.2$.

For the initial conditions, a fairly loaded network was chosen with the following values

$$l_i(0) = 30 \;, \quad i = 1, 2, 8, 9, 11, 12$$
$$l_i(0) = 70 \;, \quad i = 3, 4, 5$$
$$l_i(0) = 40 \;, \quad i = 6, 7$$
$$l_{10}(0) = 20 \,.$$

The optimization problem for this oversaturated network was solved on an IBM 370/165 digital computer using the method described. Convergence to the optimum took place in 153 iterations which required 2.73 min to execute. Table 6.2 gives the optimal control sequences and figure 6.8 the resulting optimal state trajectories of the system. Note that the queues

k	0	1	2
u_1	0.2	0.2	0.2
u_2	0.7	0.68	0.61
u_3	0.61	0.62	0.63

Table 6.2 - Optimal control sequence for the network example

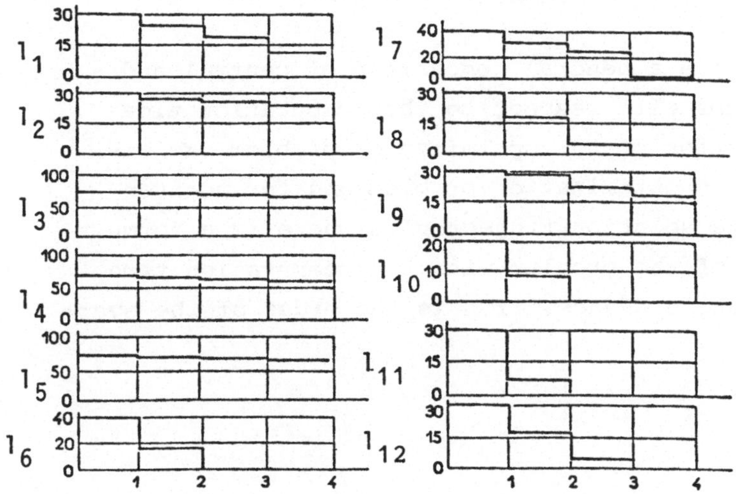

Figure 6.8
Optimal state trajectories for the London network.

l_3, l_4, l_5 are hardly reduced. The reason for this are the high inflows d_3, d_4, d_5 compared to the corresponding saturation flows s_3, s_4, s_5 so that even though the control u_1 is such that the maximum permitted green is provided for these queues, the saturation flow is too small for the queues to be dissipated.

6.3 Implementation on a multi-microcomputer system

In this section an extensive investigation of the efficiency of the dynamic decomposition algorithm based on the interaction prediction principle will be provided. The interaction prediction approach seems to be superior to other proposed algorithms for freeway traffic control purposes, as was already shown in section 6.2a). Section 6.3a) gives some general considerations regarding computation time, storage space and communication data needed for both the central and decomposed treatment of the problem. Section 6.3b) presents the results of a preliminary investigation of the problem on a main-frame computer. Section 6.3c) describes the structure of the multi-microprocessor system used and the implementation of the optimal freeway traffic control problem. The main statements of section 6.3 have been taken from /101/.

a) Computational effort

Computation time

First, we will provide a general comparison of central and decomposed algorithms with respect to the computation time needed for the solution of the optimization problem. We assume that for a given optimization problem and for a given initial guess of the solution trajectories there is a formula relating the order of the problem with the computation time needed for its solution. Hence, if n is the order of the overall problem

$$t_c = \psi(n) \tag{6.60}$$

should be the computation time needed for the central solution with ψ a nonlinear function depending upon the problem under consideration. The overall problem is now decomposed into N independent subproblems of orders n_i, $i = 1, \ldots, N$, according to section 6.1. The computation time needed for the solution of each subproblem at the first second-level (coordination) iteration is also given by the nonlinear function ψ :

$$t_i^1 = \psi(n_i) \,. \tag{6.61}$$

Thus, if the decomposed algorithm is implemented on a single computer, the computation time needed for the first second-level iteration is given by

$$t_{ds}^1 = \sum_{i=1}^{N} \psi(n_i) \,. \tag{6.62}$$

If a multiprocessor system is used, we have

$$t_{dm}^1 = \max \; \{\psi(n_1), \ldots, \psi(n_N)\} \tag{6.63}$$

for the computation time of the first iteration.

For each subsequent second-level iteration we usually have an improved initial guess of the solution trajectories and hence shorter computation times. In order to consider this fact, we approximate the computation time needed for the L-th second-level iteration by use of the relationship

$$t_i^L = t_i^1 \cdot \alpha^{L-1}, \quad L = 1, \ldots, \Lambda \tag{6.64}$$

with $0 < \alpha \leq 1$ depending on the specific optimization problem under consideration. α^{L-1} in eqn. (6.64) denotes α power $L-1$. Λ is the total number of second-level iterations.

If the time needed for the coordination task and the transfer of coordination data is assumed negligible, the total computation time for the decomposed algorithm is given by

$$t_{ds} = \sum_{L=0}^{\Lambda-1} \sum_{i=1}^{N} \psi(n_i) \,.\, \alpha^L \tag{6.65}$$

for a single computer, and

$$t_{dm} = \sum_{L=0}^{\Lambda-1} \max \{\psi(n_1), \ldots, \psi(n_N)\} \alpha^L \tag{6.66}$$

for a multicomputer system.
For example, if $\psi = n^2$ and $n_i = n/N$, $i = 1,\ldots,N$, we obtain the ratios

$$t_c/t_{ds} = N/ \sum_{L=0}^{\Lambda-1} \alpha^L \geq N/\Lambda \tag{6.67}$$

$$t_c/t_{dm} = N^2/ \sum_{L=0}^{\Lambda-1} \alpha^L \geq N^2/\Lambda \tag{6.68}$$

which means that a reduction of the total computation time is possible even on a single computer, if the number of the subsystems is greater than the number of the second-level iterations of the decomposed algorithm. If a multiprocessor system is used, the relation $N^2 > L$ provides a sufficient condition for a computation time reduction. If ψ is a linear function, above ratios become

$$t_c/t_{ds} = 1/ \sum_{L=0}^{\Lambda-1} \alpha^L \leq 1 \tag{6.69}$$

$$t_c/t_{dm} = N/ \sum_{L=0}^{\Lambda-1} \alpha^L \geq \frac{N}{\Lambda} \,. \tag{6.70}$$

which means that a computation time reduction is only possible if a multiprocessor system is used.

Storage space requirements

Let P denote the storage space needed for the overall optimization program and P_i the storage space of the subprograms. In addition, two n-dimensional trajectories ($\underline{x}$, $\underline{\lambda}$), one m-dimensional trajectory ($\underline{u}$) and one r-dimensional trajectory ($\underline{\mu}$) must be stored for the central solution. We obtain for the storage space

$$S_c = P + (2n + m + r) \quad K \; S \tag{6.71}$$

where S is the storage space needed for one variable.

If we assume χ_i interaction variables acting on the subystem i and ν_i interaction terms including local variables of the i-th subsystem, the storage space for one subproblem in the case of the prediction principle is given by

$$S_{di} = P_i + (2n_i + m_i + r_i + \chi_i + \nu_i) \cdot K \; S, \tag{6.72}$$

since we have χ_i interaction variables $\underline{\pi}_i$ and ν_i Lagrange multipliers $\underline{\beta}_i$ according to eqn. (6.19). For the coordinator we assume negligible program storage space and obtain

$$S_{co} = K \; S \sum_{i=1}^{N} (\chi_i + \nu_i) \; . \tag{6.73}$$

Thus, the total storage space in the case of a multiprocessor system is given by

$$S_{dm} = S_{co} + \sum_{i=1}^{N} S_{di} = \sum_{i=1}^{N} P_i + K \; S \; [2n+m+r+2 \sum_{i=1}^{N} (\chi_i + \nu_i)]$$

and with eqn. (6.71)

$$S_{dm} = S_c + \sum_{i=1}^{N} P_i - P + 2 \; K \; S \sum_{i=1}^{N} (\chi_i + \nu_i) . \tag{6.74}$$

Since

$$\sum_{i=1}^{N} (\chi_i + \nu_i) = 2 \sum_{i=1}^{N} \chi_i = 2 \; \chi \tag{6.75}$$

with χ the total number of interaction variables, we obtain

$$S_{dm} = S_c + \sum_{i=1}^{N} P_i - P + 4\ K\ S\ \chi . \qquad (6.76)$$

Of course, in the case of a single computer, every variable is stored only once and we have

$$S_{ds} = S_c + 2\ K\ S\ x. \qquad (6.77)$$

Eqns. (6.76) and (6.77) make clear, that a rise of the overall storage space will be the consequence of application of hierarchical optimization algorithms. However, the total storage space can be distributed on several independent computer units constituting a multicomputer system. In this case, a substantial improvement of the reliability of the overall computation structure can be achieved, if some redudant computer units are provided (see /95/ for details).

Communication Data

If a multiprocessor system is used for the solution of the decomposed optimization problem, communication channels are needed for the exchange of coordination data between the several subsystems and the coordinator unit as indicated in figure 6.3. At each second-level iteration each subsystem recieves the interaction vector $\underline{\pi}_i$ (χ_i components) and a subset of the Lagrange-multipliers vector $\underline{\beta}$ (ν_i components) and submits ν_i components of the local state vector $\underline{x}_i$. Thus, the total number of communication data during the optimization procedure becomes

$$C = \Lambda\ K\ S \sum_{i=1}^{N} (\chi_i + 2\nu_i) . \qquad (6.78)$$

Eqn.(6.78) shows that C will be low, if the couplings between the subsystems are sparse. In addition, a fast transmission of communication data between the coordinator and the independent computer units will make the according transmission times negligible compared to the computation times needed for the solution of the subproblems.

b) Solution on a single computer

Consider a freeway traffic control problem similar to the one of section 6.2a) but having only four 1 km-long segments in each freeway section. As a standard case five freeway sections will be considered.

Two different initial conditions are considered

(i) Uncongested traffic, characterized by the initial values

$$c_i^j(0) = 20, \quad v_i^j(0) = 100$$

$$j=1,\ldots,4; \quad i=1,\ldots,5$$

(ii) Congested traffic, characterized by the initial values

$$c_i^j(0) = \begin{cases} 120 & \text{for } i=3,\ j=3 \\ 80 & \text{for } i=3,\ j=2,4 \\ 50 & \text{else} \end{cases}$$

$$v_i^j(0) = \begin{cases} 15 & \text{for } i=3,\ j=3 \\ 30 & \text{for } i=3,\ j=2,4 \\ 60 & \text{else .} \end{cases}$$

The on-ramp demands, exit rates, nominal state and control values and optimization horizon are identical to those of section 6.2a).

Before implementing the decentralized optimization structure on a multiprocessor system, some preliminary results can be obtained by solution of the overall optimization problem in its central and decomposed form of a single computer.

With respect to the computation time, this preliminary investigation should provide an answer to the question whether a reduction of computation time can be expected by an implementation on a multiprocessor system. For this purpose, the relation between the order of the optimization problem and the computation time needed for its solution has been studied first. The central optimal control problem was solved on a main-frame computer (CYBER 175) for a freeway with 2,3,... 7 sections corresponding to the orders 18,27,...,63. The results, shown in figure 6.9, indicate a roughly linear dependence between computation time and problem order. Thus, in view of the results of section 6.3a) it becomes apparent that no computation time reduction can be expected by application of the decomposition methods on a single computer.

Now we ask the question: How can we get an estimate of the computation time needed for the solution of the decomposed algorithm on a multiprocessor system using only a single main-frame computer? In order to answer this question the decomposed optimal control problem has been solved for N= 2,3,...7 on the same main-frame computer as for the central case. It was found, that the number of second-level iterations is more or less independent of the number N of the subsystems included in the problem formulation. Hence, comparing the computation times for the decomposed solution with those of the central solution a computation time reduction can be expected for a sufficiently high order of the overall problem, if a multiprocessor system is used.

An estimate of the computation time t_{dm} can be made by use of the results obtained on a single computer. Let c_i^L denote the number of first level iterations of the i-th subsystem during the L-th second-level iteration. Then, the total number of first-level iterations is given by

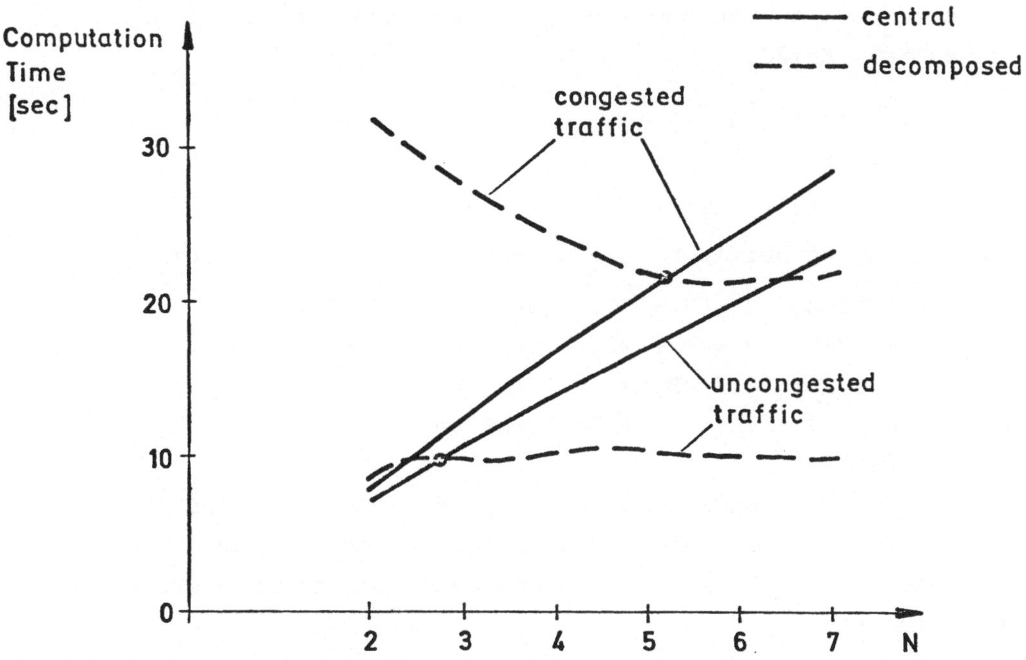

Figure 6.9 - Computation time for central and decomposed solution and its dependence on the number of freeway sections

$$Z = \sum_{L=1}^{\Lambda} \sum_{i=1}^{N} c_i^L . \tag{6.79}$$

The average computation time for each first-level iteration is equal to the ratio t_{ds}/Z. Hence, a good estimate of the multiprocessor computation time can be found by the formula

$$t_{dm} \approx \frac{t_{ds}}{Z} \sum_{L=1}^{\Lambda} \max \{c_1^L, \ldots, c_N^L\} . \tag{6.80}$$

Figure 6.9 shows the computation times t_{dm} specified by use of eqn. (6.80) for N=2,3,...,7. A computation time reduction of the multiprocessor system solution compared to the central solution is achieved, if $N > 3$ for uncongested traffic, and $N > 6$ for congested traffic, as shown in figure 6.9.

With respect to the storage space, we have for our standard case (N=5, K=120):

$P_1 = P_2 = \ldots = P_5 = P$
$n = 45$, $m=5$, $r = 10$

$n_i=9$, $m_i=1$, $r_i=2$, $k_i=3$, $\nu_i=3$, $i=1,\ldots,5$

and hence, we obtain by use of eqns. (6.71)-(6.74) for S=4 Bytes

$S_c = P + 12600 \ S = P + 50.4$ kByte
$S_{di} = P + 3240 \ S = P + 12.96$ kByte
$S_{co} = 3600 \ S = 14.4$ kByte
$S_{dm} = 5P + 19440 \ S = 5P + 77.76$ kByte.

The value of S_c indicates that one of the larger minicomputers is required for the central solution, whilst the amounts of data S_{di} and S_{co} can be accommodated in common microcomputer memories.

The amount of communication data at each second-level iteration is obtained by use of eqn. (6.78):

$C = 5400 \ S = 21.6$ kByte.

Our preliminary results with respect to the freeway traffic control problem can now be summarized as follows :

(i) A computation time reduction can be achieved on a multiprocessor system, if the order of the optimal control problem is sufficiently high.

(ii) The independent subproblems can be implemented on micro computer systems.

(iii) The communication data rate is low.

These results provide a justification for the implementation of our problem on a multi-microcomputer system.

c) The multi-microcomputer system

The multi-processor system used consists of three microcomputer systems and a minicomputer of the type INTERDATA M70. They are connected through a Universal-Interface-Module (UIM) in a star configuration (figure 6.10). Three of the described subproblems of the freeway traffic control problem are implemented (ASSEMBLER) and solved in the three microcomputers. The minicomputer treats the remaining two subproblems and the coordination task. Since the minicomputer is more than twice quicker than the microcomputers, the solution of the independent subproblems can be viewed as being parallel.

Every microcomputer system contains a Z80 microprocessor coupled with an AM 9511 Arithmetic Processor Unit and 16k RAM storage space. For the communication of the microcomputers with the minicomputer the direct access memory method has been chosen enabling the transmission of 10 kByte per second.

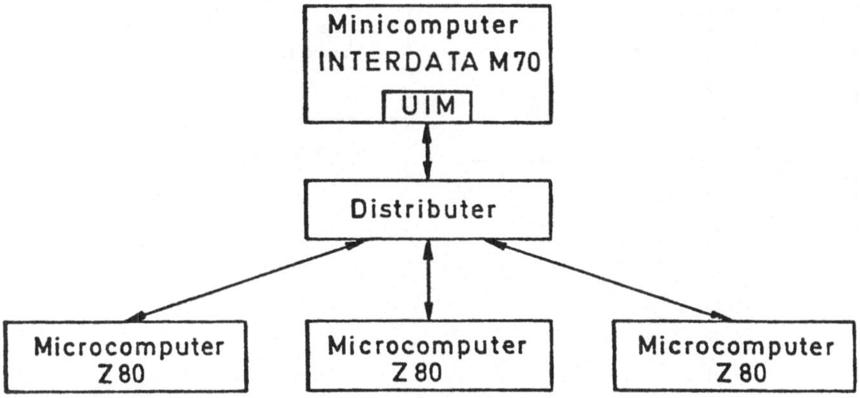

Figure 6.10 - The multiprocessor system.

The synchronization of the actions of the various computers is handled by the minicomputer as shown on the flow diagram of figure 6.11. At the beginning of a second-level iteration the minicomputer sets the value of a flag-variable of each microcomputer to one and then solves its own subproblems. When a microcomputer finishes its task, its flag-variable is set to zero. The minicomputer reads periodically the value of the flag of the microcomputers, and if they are all zero, he proceeds to the coordination task and so on.

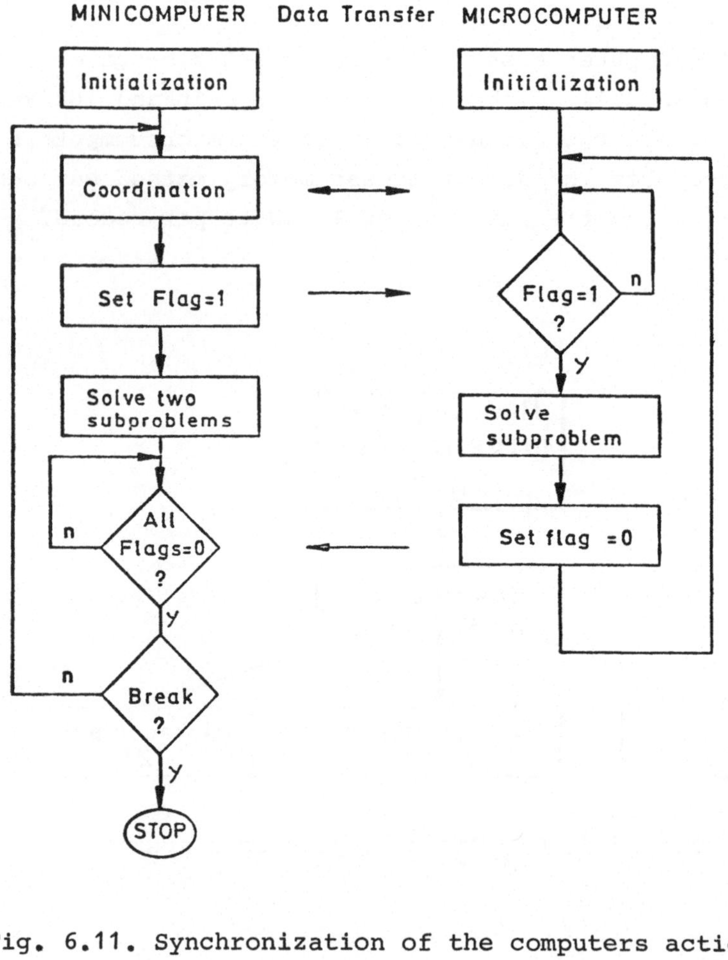

Fig. 6.11. Synchronization of the computers actions.

Three equal subproblems each taking 16 kBytes of storage space have been implemented on the three microcomputers, which can be programmed and tested independently by appropriate I/O units. The minicomputer programs have been written in FORTRAN 5. The two subproblems and the coordinator task needed more than 64 kByte of the minicomputer's main memory. Hence a disc memory together with overlay techniques had to be used.

The computation time for one first-level iteration of the microcomputers has been found to be 15 seconds. A considerable part of this time was consumed for the calculation of the complicated formula (5.8). By use of a simplified table-form approximation, the computation time for one first-level iteration was reduced to 9 seconds. The data transfer at each second-level iteration for five subsystems amounts to 20 kBytes corresponding to 2 seconds of transmission time. Hence, the data transfer times can indeed be viewed as negligible in case of the prediction principle algorithms.

The computation time needed for the solution of the overall problem has been 12 minutes for the uncongested and 22 minutes for the congested traffic. The results have been equal to those obtained on the main-frame computer.

Of course, 22 minutes computation time is still too long compared to the 30 minutes long optimization horizon considered. However, a further reduction of computation time seems to be possible in view of the most recent developments of microprocessor technology. As already mentioned, a reduction is desired up to a degree allowing on-line computation of the optimal control strategy. With regard to the optimization horizon considered in our problem formulation, a computation time in the order of 1 minute seems to be short enough for an on-line control structure.

7. THE MULTILAYER APPROACH TO THE SOLUTION OF TRAFFIC CONTROL PROBLEMS.

Most of the developed nonlinear optimal control theory is concerned with the mathematical problem of determining optimum control for systems described by mathematical models subject to various constraints. Evaluation of control system structures considering real-life aspects, such as real-time requirements, effects of disturbances, model inaccuracy, tradeoff between suboptimal performance and implementation cost etc., has found little attention in the control theoretical literature /102, 103/. The importance of these topics for the evaluation of a control system becomes evident when dealing with complex, large-scale systems.

It is the purpose of this chapter to present a multilayer control scheme /104-106/ for suboptimal control of large-scale nonlinear plants taking into account the actual process conditions in all levels of decision making. The basic philosophy of multilayer control hierarchies consists in a vertical decomposition of the overall control system into simpler subproblems of a form readily solved and implemented by available techniques.

The multilayer control structure described in this paper has been developed as a solution of the freeway traffic control problem. Nontheless it can be successfully applied to other technical and nontechnical control problems /107-109/.

7.1 General notions

a) General problem formulation and possible solution structures

Consider the control system shown in figure 7.1, $\underline{u} \in R^m$

is the input vector, $\underline{x} \varepsilon R^n$ is the state vector, $\underline{z} \varepsilon R^q$ are measurable, low frequency disturbances, which cannot be manipulated by the control system and $\underline{\zeta} \varepsilon R^p$ are disturbances for which no a-priori information is available. A quite general optimal control problem P7 can be formulated[1]):

On the basis of the actual and past values of the state variables $\underline{x}$ and slow disturbances $\underline{z}$ find the sequence of input variables $\underline{u}^*(k)$, $k = 0, \ldots, K-1$, minimizing the cost functional

$$J = \sum_{k=0}^{K-1} \phi \, [\underline{x}(k), \underline{u}(k), \underline{z}(k)] \tag{7.1}$$

subject to the mathematical model equation

$$\underline{x}(k+1) = f \, [\underline{x}(k), \underline{u}(k), \underline{z}(k), \underline{\zeta}(k)] \; ; \; \underline{x}(0) = \underline{x}_o \tag{7.2}$$

and the inequality constraints

$$\underline{h} \, [\underline{x}(k), \underline{u}(k), \underline{z}(k)] \underline{\times} \underline{0} \tag{7.3}$$

holding for $k=0,\ldots,K-1$.

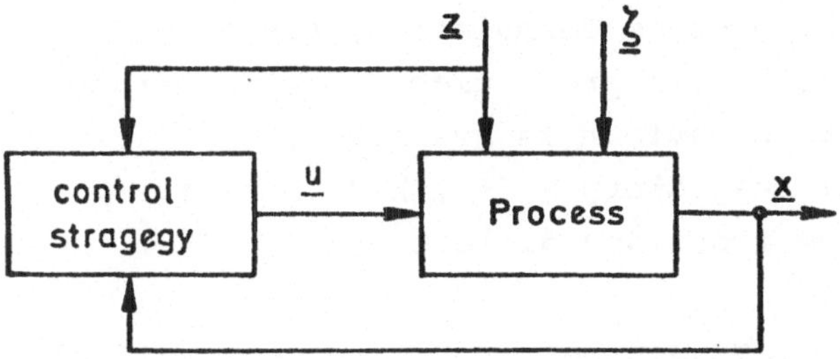

Figure 7.1
A control system.

---

(1) The estimation problem is not included in this problem formulation. Accurate measurements (or estimates) of $\underline{x}$ and $\underline{z}$ are assumed.

Existing model inaccuracies are being viewed as additional unknown disturbances included in the vector $\underline{\zeta}$.

Closed-loop control

From the viewpoint of implementation, closed-loop control strategies represent the simplest solution of an optimal control problem. Development of these strategies results in an analytical feedback law of the form

$$\underline{u}^*(k) = \underline{L}\,[\underline{x}(k), \underline{z}(k), k] \quad ; \; k = 0,\ldots, K-1 \qquad (7.4)$$

that is, the optimal control sequence can be computed on-line as an explicit function of the actual values of $\underline{x}$ and $\underline{z}$ and the time k. Unfortunately, evaluation of a feedback law requires the explicit solution of the Hamilton-Jacobi-Bellman equation, which is feasible only for specific problems (e.g. for LQ-problems in absence of inequality constraints). Different control structures must be chosen in all other cases.

Open-loop control

Open-loop control strategies can be evaluated off-line by means of the Pontryagin's Maximum Principle, if the following two assumptions are met :

(i) $\underline{\zeta}$ is dropped in the problem formulation, since no deterministic or stochastic properties, i.e. no mathematical model of its behaviour is available.

(ii) A predicted nominal trajectory $\underline{z}_N(k)$, $k=0,\ldots,K-1$, is provided for the low frequency disturbances.

As a consequence of these assumptions, $\underline{x}(k)$ can be specified by all prior values of $\underline{u}$, $\underline{z}_N$ and its initial condition $\underline{x}_0$ in a unique way. Thus, open-loop optimal control $\underline{u}_0^*(k)$ is evaluated on the basis of $\underline{z}_N(k)$ and $\underline{x}_0$ only :

$$\underline{u}_o^*(k) = \underline{L}_o \left[\underline{z}_N(0), \ldots, \underline{z}_N(K-1), \underline{x}_o\right] \quad ; \; k=0,\ldots,K-1. \qquad (7.5)$$

Remarks: 1) Solution of the open-loop optimal control problem is obtained through solution of a TPBVP resulting from necessary conditions for optimality /81/. Hence, unlike equation (7.4), a procedural rather than an analytical relationship is formally expressed in equation (7.5).

2) Solution of a TPBVP is not a trivial task for a large-scale system. Time consuming and storage space extensive, iterative algorithms must be used. Good convergence properties can only be guaranteed for special cases.

3) If existence of singular arcs is not excluded, extreme difficulties may arise when dealing with large-scale plants (see section 5.1.e).

4) As a by-product of the evaluation of $\underline{u}_o^*$, an open-loop optimal state trajectory $\underline{x}_o^*(k)$, k=1,..., K, is obtained.

5) Application of open-loop control implies exitation of the physical system with the stored open-loop optimal trajectory $\underline{u}_o^*$.

6) Actual measurements of $\underline{x}$ and $\underline{z}$ are not used in this control structure (figure 7.2).

Open-loop control structures suffer from following drawbacks:

1) Extensive off-line calculations are needed for the evaluation of the optimal trajectories.

2) Open-loop optimal trajectories are more or less different than the desired solution of problem P2, i.e. $\underline{x}^* \neq \underline{x}_o^*$ and $\underline{u}^* \neq \underline{u}_o^*$, since

- Future values of disturbances are generally not exact known, i.e. $\underline{z}(k) \neq \underline{z}_N(k)$,
- Unexpected disturbance $\underline{\zeta}$ may occur,
- Model inaccuracies are present.

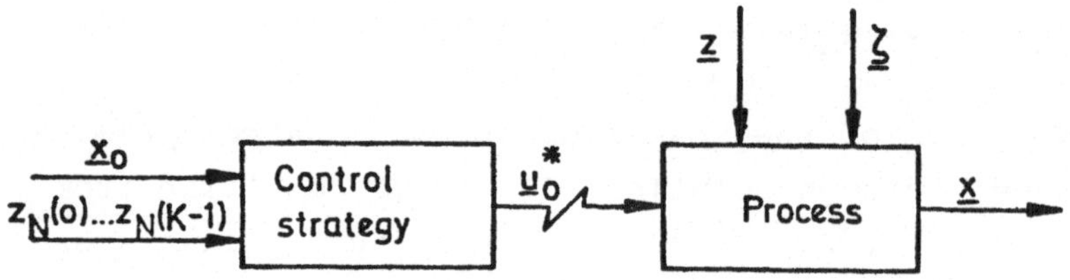

Figure 7.2 - Open-loop control.

3) For the same reasons, exitation of the real process with $\underline{u}_o^*(k)$ will result in real life state trajectories, which are different than the open-loop optimal ones, i.e. $\underline{x} \neq \underline{x}_o^*$.

Two-stage control structure

If only small deviations of the disturbances from their nominal values are expected, a linearisation of the model equation around the open-loop optimal trajectories may be admissible /110/

$$\delta\underline{x}(k+1) = A\ \delta\underline{x}(k) + B\ \delta\underline{u}(k) + C\ \delta\underline{z}(k) \quad ; \quad \delta\underline{x}(0) = 0 \tag{7.6}$$

where $\delta\underline{x}=\underline{x}-\underline{x}_o^*$, $\delta\underline{u}=\underline{u}-\underline{u}_o^*$, $\delta\underline{z}=\underline{z}-\underline{z}_N$ and

$$A = \partial\underline{f}/\partial\underline{x}|_o \qquad B = \partial\underline{f}/\partial\underline{u}|_o \qquad C = \partial\underline{f}/\partial\underline{z}|_o \tag{7.7}$$

are time varying matrices.

In order to keep the actual plant state $\underline{x}(k)$ near its open-loop optimal values $\underline{x}_o^*(k)$, minimization of the following quadratic functional can be required.

$$J_Q = 1/2 \sum_{k=0}^{K-1} \{\delta\underline{x}(k)^T Q\ \delta\underline{x}(k) + \delta\underline{u}(k)^T R\ \delta\underline{u}(k)\} + $$
$$+ 1/2\ \delta\underline{x}(K)^T Q_K\ \delta\underline{x}(K) \tag{7.8}$$

where Q, Q_K, R ≥ 0 are appropriately selected weighting matrices. Thus, in addition to open-loop optimization, the following first-stage problem P8 can be formulated:

> On the basis of actual and past values of $\underline{x}$ and $\underline{z}$, the open-loop optimal trajectories $\underline{x}_o^*$ and $\underline{u}_o^*$ and the nominal trajectory $\underline{z}_N(k)$ find the sequence $\delta\underline{u}^*(k)$, k=0,...,K-1, minimizing the cost functional (7.8) subject to equations (7.3), (7.6).

If small constant deviations of the disturbances are present, $\delta\underline{z} \approx \text{const.} \neq \underline{0}$, an optimal feedback law can be derived in absence of inequality constraints (7.3):

$$\delta\underline{u}^o(k) = -L(k)\ \delta\underline{x}(k) - G(k)\ \delta\underline{z}(k) \tag{7.9}$$

where L, G are time varying control gain matrices obtained by solution of a matrix difference equation corresponding to the matrix Riccati equation for continuous systems /42/. If $\delta\underline{u}^o(k)$ violates the inequality constraints (7.3), it should be appropriately changed at each time period by some heuristic procedure. The structure of the two-stage approach is shown in figure 7.3.

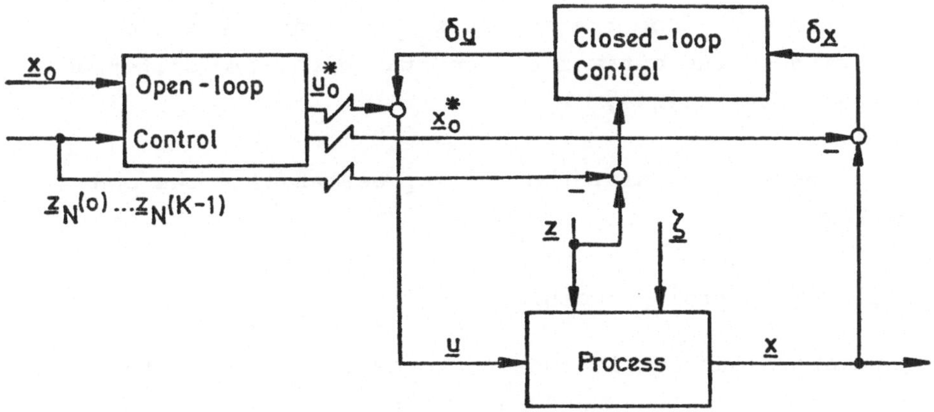

Figure 7.3-Two-stage control structure.

Let us discuss the properties of this structure compared to open-loop strategies:

1) Implementation cost has been increased, since a) additional off-line calculations are performed for the control gain matrices L and G and b) closed-loop control with storage capabilities must be provided.

2) Closed-loop control has no influence on the results of the optimization procedure. Hence, open-loop optimal trajectories will still be different than the desired solution of P7, i.e. $\underline{x}^* \neq \underline{x}_o^*$ and $\underline{u}^* \neq \underline{u}_o^*$.

3) Selection of the quadratic performance functional (7.8) drives real-life state trajectories near the open-loop optimal ones, i.e. $\underline{x} \approx \underline{x}_o^*$. Hence, it can be guaranteed that the control performance calculated for the open-loop control strategy will be achieved even in presence of small disturbances and slight model inaccuracies.

The drawbacks of the two-stage approach are similar to those of the open-loop control:

1) If the nominal trajectories of the disturbances are not accurate, the open-loop trajectory $\underline{x}_o^*$ will not be a good approximation of the optimal one and the quadratic criterion (7.8) makes little sense.

2) Strong unexepected disturbances may destabilize the control system.

b) Multilayer_control_structure

Letus summarize our conclusions up to here. A mathematical optimal control problem has been formulated as an approximation of the real optimal control problem. Since no general closed-loop solution can be evaluated, open-loop structures approximating the optimal solution must be chosen. Some drawbacks of the open-loop structure can be partially compensated by introducing a first stage LQ-controller, However, the two-stage approach is still nonsatisfactory for processes with inaccurate prediction of slow disturbances and occurence of strong, unexpected disturbances. An improvement can be achieved, if feedback information is provided to the open-loop optimization procedure which has to be periodically solved on-line.For a large-scale system, this becomes a feasible task, only if a simplified version of the optimization problem is considered. The loss of performance caused by the simplification is partially removed by appropriate design of two additional control layers resulting in the multilayer structure of figure 7.4.

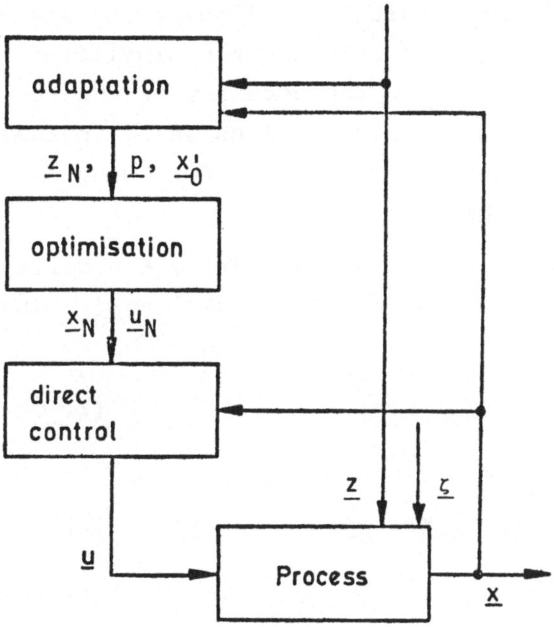

Figure 7.4
A multilevel control structure.

Optimization layer

Occurrence of singular subarcs, low convergence rates, real-time requirements and constrained implementation funds make the periodical on-line solution of an open-loop optimization problem for a large-scale system an extremely difficult task.

Simplification of optimization problem P7 can be achieved by use of well-known aggregation techniques /111/, separation of slow and fast dynamics /112/ or other less general approaches depending on the specific properties of the process under control.
Determination of an optimal steady-state point can be the task of the optimization layer if the regulating control actions of the direct control layer are capable of maintaining the process reasonably close to the determined steady-state.

In any case, some kind of feedback information from the process to the optimization layer should be available, for example in form of a parameter vector $\underline{p}$ entering the simplified optimization problem and influencing the accuracy of the results according to the current state conditions as is indicated in figure 7.4. Thus, we obtain formally the following simplified optimization problem P9:

Given some predicted slow disturbances $\underline{z}_N(k)$, $k = 0,\ldots,K'$, the parameter vector $\underline{p}$ and the initial condition $\underline{x}'_o$, minimize the cost functional

$$J' = \sum_{k=0}^{K'-1} \phi'[\underline{x}'(k),\ \underline{u}(k),\ \underline{z}_N(k),\underline{p}] \tag{7.10}$$

subject to the constraints

$$\underline{x}'(k+1) = f'[\underline{x}'(k),\ \underline{u}(k),\ \underline{z}_N(k),\ \underline{p}];\ \underline{x}'(o) = \underline{x}'_o \tag{7.11}$$

$$\underline{h}'[\underline{x}'(k),\underline{u}(k),\underline{z}_N(k),\underline{p}] \geq \underline{0} \tag{7.12}$$

$k = 0,\ldots, K'-1.$

where $\dim(\underline{x}) > \dim(\underline{x}')$ or/and $K' < K$. Equations (7.10)-(7.12) represent a simplified version of equations (7.1)-(7.3). An inverse relation giving the original state vector $\underline{x}$ in terms of the reduced state vector $\underline{x}'$ is assumed known

$$[\underline{x}^T(1)\ \ldots\ \underline{x}^T(K)] = \underline{g}[\underline{x}'^T(0)\ldots\underline{x}'^T(K')]. \tag{7.13}$$

The input and state variable trajectories obtained by solution of problem P9 and application of eqn. (7.13) will be refered in the following as the nominal trajectories and will be denoted by $\underline{u}_N(k)$, $\underline{x}_N(k)$, $k = 0,\ldots,K-1$.

Direct control layer

Direct control layer must translate the decisions of optimization layer into real control actions and suppress the effects of slight model inaccuracies and unexpected disturbances. Simplifications introduced in the optimization layer must be consistent with the design of the direct control layer in order to get a satisfactory overall control behaviour /103/.

The linear quadratic problem P8 of section 7.1a) resulting in an analytical feedback law can be chosen as the task of the direct control layer. In that case, the nominal trajectories $\underline{x}_N(k)$, $\underline{u}_N(k)$ determined in the optimization layer are considered as desired trajectories in the quadratic performance criterion (7.8). If the plant under consideration can be subdivided into a number of interconnected subplants, wellknown methods /113,114/ for the development of decentralized or overlapping feedback laws can be applied leading to better reliability properties and lower implementation cost.

The two-layer structure developed up to here is similar to the two-stage structure of section 7.1a). Let us recall its main drawbacks:

1) Actual measurements of slow disturbances are not used in the optimization layer. As a result, determination of nominal trajectories (or nominal set points) is inaccurate, if predicted disturbance trajectories are inaccurate.

2) Unexpected, strong disturbances may drive the process state to a region for which the problem simplifications met in the optimization layer are not valid leading to destabilization of the control system.

3) As a result of the simplifications introduced, nominal trajectories will be suboptimal even if inaccuracies 1) and 2) are not present. However, simplifications leading to a substantial reduction of the computational effort enable an on-line solution of the optimization problem. Thus, introduction of an additional adaptation layer providing the optimization layer with some feedback and feedforward information will lead to partial elimination of the three drawbacks.

Adaptation Layer

The task of the adaptation layer is the specification of the predicted values of the disturbances $\underline{z}_N$ and the model parameters $\underline{p}$. The model parameter vector is chosen from a given set of discrete values according to the current process conditions so as to guarantee that a reasonable approximation of the optimal solution will be achieved in spite of the simplifications of the optimization layer and the linearization in the direct control layer.

The most simple way of specifying $\underline{p}$ is according to the following rule

$$\underline{x}(k) \in X_i \Rightarrow \underline{p} = \underline{p}_i \tag{7.14}$$

where X_i are subsets of the state space, such that

$$\bigcup_{i=1}^{P} X_i = R^n \quad \text{and} \quad X_i \cap X_j = \emptyset \quad \forall i,j \in [1,P].$$

$\underline{p}_i$ have prespecified values such that in the corresponding state space region

i) Open-loop optimal control is approximated

ii) Stability of the control system is guaranteed.

For example, consider the specification of a set point $(\bar{\underline{x}}_N, \bar{\underline{u}}_N)$ as the task of the optimization layer for a time invariant system with known constant disturbances $\underline{z}=\bar{\underline{z}}_N$. Solution of a LQ problem with infinite time horizont in the direct controllayer results in the feedback law

$$\delta\underline{u}(k) = - L\ \delta\underline{x}(k) \tag{7.15}$$

where L is a time invariant feedback gain matrix. The control system is then described by the difference equation

$$\underline{x}(k+1) = \underline{f}\ [\underline{x}(k),\ \bar{\underline{u}}_N - L\ [\underline{x}(k) - \bar{\underline{x}}_N],\ \bar{\underline{z}}_N\ ,\ \underline{\zeta}(k)]\ . \tag{7.16}$$

For $\underline{\zeta}(k) \equiv \underline{0}$, the region of attraction of the nominal state point (figure 7.5) depends on the nominal values $\bar{\underline{x}}_N$, $\bar{\underline{u}}_N$, $\bar{\underline{z}}_N$, i.e.

$$\Lambda = \Lambda(\bar{\underline{x}}_N,\ \bar{\underline{u}}_N,\ \bar{\underline{z}}_N).$$

Since $\bar{\underline{x}}_N$, $\bar{\underline{u}}_N$ represent the solution of optimization problem P9, we can write

$$\Lambda = \Lambda(\underline{p},\ \bar{\underline{z}}_N).$$

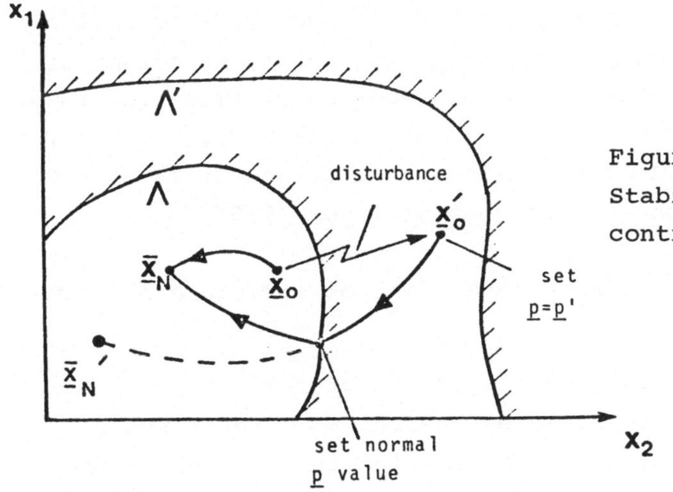

Figure 7.5
Stabilization of the control system.

Hence, any time the current process state $\underline{x}$ is driven outside Λ by some strong disturbance, an appropriate change of $\underline{p}$ as described above, can lead to a new nominal state point $(\bar{\underline{x}}_N', \bar{\underline{u}}_N')$ and a new region of attraction Λ' including the current state. In that way, the current state can be driven back to the normal region Λ. When this is achieved, a normal value of $\underline{p}$ according to eqn. (7.14) is again chosen.

Predicted trajectories $\underline{z}_N$ for low frequency disturbances may be obtained from historical data. If great deviations $\|\underline{z}(k) - \underline{z}_N(k)\|$ occur, the nominal trajectories $\underline{x}_N$, $\underline{u}_N$ specified in the optimization layer become inaccurate. For this reason, predicted disturbances $\underline{z}_N(k)$ are appropriately changed in the adaptation layer and a new optimization run is started. In order to keep the frequency of new optimization runs low, simple pattern classification procedures can be used considering the extension and duration of the occuring deviations /67/. For complex processes, intelligent identification systems might be useful in this context /115/. On the other hand, periodical solution of the optimization problem with changed predicted trajectories and updated initial conditions leads to a partial compensation of the effect of model inaccuracies. The period of repeated optimization runs should be shorter than the optimization horizont, i.e. only a part of the specified nominal control trajectories should be really used in the direct control layer /116/.

The properties of the three layer control structure are summarized in the following :

1) We recognize the two stage approach of section II in the first two control layers. In extension, some feedforward $(\underline{z}_N)$ and feedback $(\underline{p}, \underline{x}(0))$ information is provided to the open-loop optimization procedure which is evaluated on-line. For large-scale systems some simplifications might be required.

2) The overall control structure is rather insensitive to variations of the low frequency disturbances $\underline{z}$, since feedforward loops are provided.

3) Unexpected disturbances driving the process state away from the nominal trajectories cannot destabilize the system, since appropriate feedback loops are provided.

4) Repeated optimization starts partially eliminate the effects of model inaccuracies.

5) The multilayer structure provides a robust, suboptimal solution of the optimal control problem P7 . Rate of suboptimality depends on the specific design decisions. Its theoretical preestimation for specific cases is a task of future research.

6) The overall control system structure permits an implementation on a distributed multicomputer system /117/. Local data processing, local state estimation and local direct control layer can be implemented on a single microcomputer near the real subplant. A medium size minicomputer evolves the central tasks of adaptation and optimization as well as input and output tasks like visualization of the current process conditions, storage of process data needed for statistic analysis etc. Since only a small part of the measured data are fed back to the adaptation layer, rather low communication data rates between the central computer and the local microcomputers are expected.

c) Benchmark problem /117/

Consider the following simple control problem corresponding to problem P7 of section 7.1.a):

On the basis of the actual and past values of $\underline{x}$ and $\underline{z}$ find the input vector $\underline{u}^*(k)$, $k = 0, \ldots, K-1$, minimizing the cost functional

$$J = \sum_{k=0}^{K-1} \{0.5\, u_1(k)^2 + 10\,[u_2(k) - 0.8]^2 - 50 x_1 x_2\} \qquad (7.17)$$

subject to

$$x_1(k+1) = x_1(k) - 0.2 x_1(k)\, x_2(k) + 0.02\, u_1(k) + z_1(k) + \zeta_1(k) \qquad (7.18)$$

$$x_2(k+1) = -0.26 x_1(k) + 0.8 x_2(k) - 0.48 x_1(k) u_2(k) + + 0.25\, u_2(k) + \zeta_2(k) \qquad (7.19)$$

$$|u_1(k)| \leq z_2(k) \qquad (7.20)$$

$$0.7 \leq u_2(k) \leq 1.2 \qquad (7.21)$$

Open -loop optimal control

Assuming

$$\zeta_1(k) \equiv \zeta_2(k) \equiv 0; \quad z_{1N}(k) \equiv 0.01; \quad z_{2N}(k) \equiv 8$$

(predicted values) we can formulate and solve by use of Pontryagin's Maximum Principle a nonlinear, dynamic optimal control problem leading to the control sequence

$$u_o^*(k), \quad k = 0, \ldots, K-1$$

according to section 7.1.a). Using open-loop optimal control for the given values of the distrubances, we get stable trajectories for initial points x_o lying inside region C in figure 7.6.

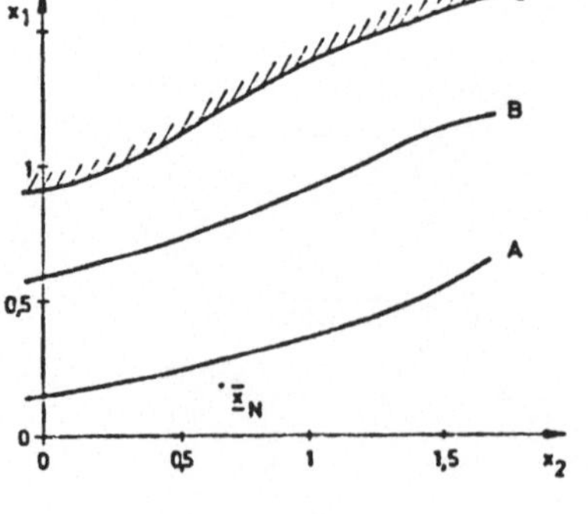

Figure 7.6
Region of attraction for several control strategies.

Multilayer Control

Optimization Layer

Consider the specification of an optimal steady-state point as the task of the optimization layer. The steady-state version of problem (7.17) - (7.21) is :

$$\text{Minimize} \qquad \bar{J} = 0.5\, \bar{u}_1^2 + 10\, (\bar{u}_2-0.8)^2 - 50\bar{x}_1\bar{x}_2 \qquad (7.22)$$

subject to

$$-0.2\bar{x}_1\bar{x}_2 + 0.02\, \bar{u}_1 + \bar{z}_{1N} = 0 \qquad (7.23)$$

$$-0.2\bar{x}_2 - 0.26\bar{x}_1 - 0.48\bar{x}_1\bar{u}_2 + 0.24\bar{u}_2 = 0 \qquad (7.24)$$

$$|\bar{u}_1| \leq \bar{z}_{2N} \qquad (7.25)$$

$$0.7 \leq \bar{u}_2 \leq 1.2\ , \qquad (7.26)$$

Solution of this steady-state optimization problem for $\bar{z}_{1N}=0.01$ and $\bar{z}_{2N} = 8$ is given by

$$\bar{u}_1^* = 0.596,\quad \bar{u}_2^* = 1.093,\quad \bar{x}_1^* = 0.167,\quad \bar{x}_2^* = 0.656\ . \qquad (7.27)$$

Applying these steady-state optimal values (without direct control layer) leads to the region of attraction A in figure 7.6. In order to enlarge the region of attraction and improve the dynamic behaviour of the system we introduce an infemal direct control layer.

Direct Control Layer

Linearization of the system equations (7.18), (7.19) around the steady-state optimal point of (7.27) leads to the matrices

$$A = \begin{bmatrix} 0.866 & -0.034 \\ -0.797 & 0.8 \end{bmatrix} \qquad B = \begin{bmatrix} 0.02 & 0 \\ 0 & 0.159 \end{bmatrix} \qquad (7.28)$$

for the linearized system according to equation (7.7). No feedforward loop for $\delta \underline{z}$ is considered. Setting

$$Q = \text{diag}\ \{10,10\}\ , \qquad R = \text{diag}\ \{0.5,\ 10\}$$

we get the feedback matrix

$$L = \begin{bmatrix} 7.48 & -1.52 \\ -0.87 & 0.34 \end{bmatrix} . \tag{7.29}$$

If the control inputs calculated according to eqn. (7.9) violate the inequality constraints (7.20), (7.21), they are being set equal to the limit values. Application of the feedback law (7.9) leads to the region of attraction B in figure 7.6.

Adaptation layer

Since $B \subset C$, an unexpected disturbance $\zeta_1(k)$ resp. $\zeta_2(k)$ could drive the closed-loop system outside its region of attraction B. In that case the adaptation layer should intervene and stabilize the closed loop system by setting, say

$$u_{1N} = p = -8, \quad \text{if} \quad \underline{x}(k) \in C-B . \tag{7.30}$$

Furthermore, the adaptation layer should change the values of $\bar{z}_{1N}$, $\bar{z}_{2N}$, if a significant deviation is being observed, say if

$$|z_1(k) - \bar{z}_{1N}| \geq 0.003$$

or

$$|z_2(k) - \bar{z}_{2N}| \geq 0.2$$

and initiate a new steady-state optimization run with corrected disturbance values.

Results

The normal case

Consider the case $\underline{z} \equiv \bar{\underline{z}}_N$, $\underline{\zeta}(k) \equiv \underline{0}$. For $\underline{x}(0) = [0.9 \quad 0.5]^T$ and K = 120, we get the results shown in figure 7.7. Multi layer control approximates the optimal control in a satisfactory way. The corresponding values of the performance functional are

$$J_o^* = -359$$

$$J_m = -311$$

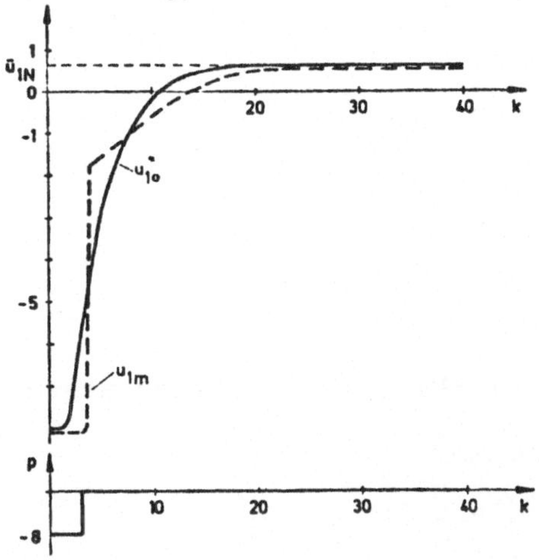

Figure 7.7
Open-loop optimal control $u_{1,0}^*$ and multilayer control $u_{1,m}$.

Occurrence of unexpected disturbances

Consider now the case

$$\zeta_1(x) = \begin{cases} 0{,}01 & \text{for} \quad k \in [10,20] \\ 0 & \text{else} \end{cases} \tag{7.31}$$

For the same initial condition. Figure 7.8 shows that the unexpected disturbance destabilizes the open-loop control system, whereas the multilayer control system remains stable.

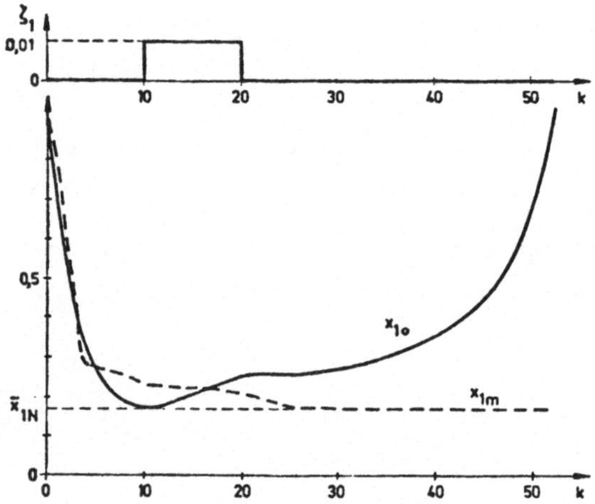

Figure 7.8

State variables $x_{1,o}$ and $x_{1,m}$ in the case of an unexpected disturbance.

<u>Occurrence of deviations δz</u>

Consider now the case

$$z_1(k) = \begin{cases} \bar{z}_{1N} & \text{for} \quad k \leq 20 \\ \bar{z}_{1N} - 0.003 & \text{for} \quad k > 20 . \end{cases} \tag{7.32}$$

This corresponds to a moving of the optimal steady-state point from (7.27) to the values

$$\bar{u}_1^* = 0.734 \ , \quad \bar{u}_2^* = 1.084 \ , \quad \bar{x}_1^* = 0.167 \ , \quad \bar{x}_2^* = 0,65 . \tag{7.33}$$

This is being detected by the adaptation layer according to the described rule and leads to the on-line specification of the new steady-state point by a new optimization run. Figure 7.9 shows the corresponding values of $x_{1,o}$ and $x_{1,m}$. In the case of multilayer control, the new optimal steady-state point is reached, whereas open-loop control converges towards a steady-state value which is not equal neither to the old nor to the new optimal point. The values of the performance functional are

$$J_o = -\ 305$$

$$J_m = -\ 320.$$

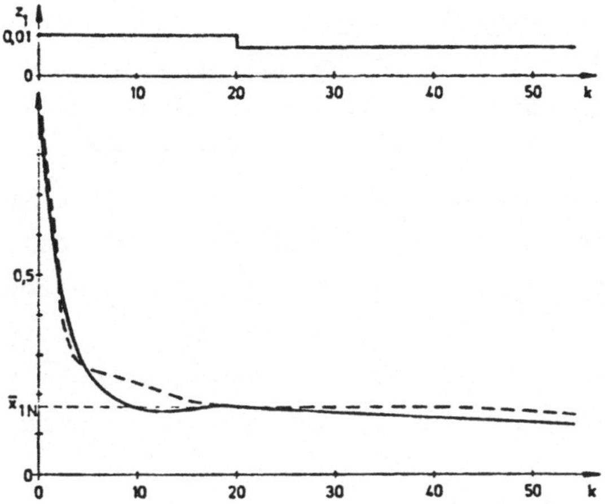

Figure 7.9 - State variables $x_{1,o}$ and $x_{1,m}$ in the case of occuring deviation $\delta \underline{z}$.

7.2 Application to traffic control

An open-loop optimal control problem P3 for freeway traffic has already been formulated and discussed in section 5.1e). It has been found that specification of optimal solution trajectories is an extremely difficult task even for an off-line treatment. For this reason, a multilayer control system will be developed in the following.

a) Simplification of the problem formulation: the optimization layer

The first question to be answered when developing the multilayer control system is how to simplify optimization problem P3 so as to make its on-line solution a feasible task and still get reasonable solution trajectories. Two alternative possibilities for the simplification of P3 depending on the geometric characteristics of the freeway under consideration are described in the following.

a1) <u>Simplified dynamic optimization</u>

Vehicle mean speed in a segment is adjusted to the actual traffic density according to the dynamic equation (3.2). Dynamic terms in (3.2) describe the reaction of vehicle drivers to changeable traffic conditions and provide the freeway traffic model with an additional degree of realism, especially in cases of congested traffic /14/. Since the main objective of the optimization layer consists in displacing traffic demand in space and time, it seems to be a reasonable simplification to neglect the dynamic terms in equation (3.2) and express v_i^j by means of the algebraic relation

$$v_i^j(k) = v_e\left[\rho_i^j(k),\ b_i(k)\right] . \tag{7.34}$$

This leads to a considerable reduction of the model order. Setting equations (3.3) and (7.34) in (3.1) we get

$$\rho_i^j(k+1) = \rho_i^j(k) + \frac{T}{\Delta_i^j}\left\{\alpha.Q\left[\rho_i^{j-1}(k),\ b_i(k)\right] + (1-2\alpha).\right.$$
$$\left. .Q\left[\rho_i^j(k),\ b_i(k)\right] - (1-\alpha).Q\left[\rho_i^{j+1}(k), b_i(k)\right]\right\} \tag{7.35}$$

where $Q(\rho,b)$ denotes the steady-state volume-density characteristic corresponding to $v_e(\rho,b)$.

Subdivision of the freeway into relatively short segments is necessary for the accurate description of congested traffic /14,29/. For uncongested traffic, densities in the various segments of a section have similar values. Hence, assuming uncongested traffic, it is reasonable to introduce an aggregate section traffic density

$$c_i = 1/n(i) . \sum_{j=1}^{n(i)} \rho_i^j . \tag{7.36}$$

Equal segment lengths in a section have been assumed. Expressing equations (3.1), (3.3), (7.34) in terms of section densities we obtain [1]

$$c_i(k+1) = c_i(k) + \frac{T}{\delta_i} \left\{ \alpha Q\left[c_{i-1}(k),\ b_i(k)\right] + (1-2\alpha)\cdot \right.$$

$$Q\left[c_i(k), b_i(k)\right] - (1-\alpha) Q\left[c_{i+1}(k),\ b_{i+1}(k)\right]$$

$$\left. + r_i(k) - s_i(k) \right\}. \qquad (7.37)$$

Note that the section length δ_i appears in eqn. (7.37) in place of the segment length Δ_i^j. Since

$$\delta_i = \sum_{j=1}^{n(i)} \Delta_i^j, \qquad (7.38)$$

the dynamic system represented by eqn. (7.37) is approximately n(i) times slower than the original one and hence a longer sample time period $T' = n(i)T$ can be selected. The performance functional (5.12) is rewritten in terms of the aggregated section density

$$T_s = T' \sum_{k=0}^{K'} \left[\underline{c}(k)^T \underline{\delta} + \underline{1}(k)^T \underline{e}\right]. \qquad (7.39)$$

In place of the O-D rates the exit rates γ_i defined by eqn. (2.16) are considered as slow disturbances. The nonconventional constraints (5.10) are changed to

$$0.7 \leq b_i(k) \leq 1.0 \qquad (7.40)$$

whilst constraints (5.11) are dropped from the problem formmulation. Appropriate modification of the obtained optimal input trajectory to fulfill constraints (5.10) and (5.11) is then an easy task.

---

(1) Equation (7.37) can also be obtained through a simplified linear form of formula (7.35) by straightforward application of linear aggregation methods /111/.

Let us now consider the presence of a congestion in the j-th segment of the i-th section. In that case, aggregation by means of equation (7.36) would disperse the overcritical density of segment j all over the section resulting in a totally wrong description of freeway traffic behaviour and hence, in inadequate control actions. The question now arising is: How should the simplified model of eqn. (7.37) be altered in order to lead to reasonable results in case of congestion?

To answer this question we consider the congested traffic situation shown in figure 7.10. It is a well-known fact that traffic volume q_{in} entering the congested segment mainly depends on the upstream traffic state, whilst volume q_{out} leaving the congested segment depends only on the congestion density (and not on the downstream traffic state). For these reasons it is firstly necessary to introduce an additional continuity equation for the congested segment

$$\rho_i^j(k+1) = \rho_i^j(k) + \frac{T}{\Delta_i^j} \cdot [q_{in}(k) - q_{out}(k)] \qquad (7.41)$$

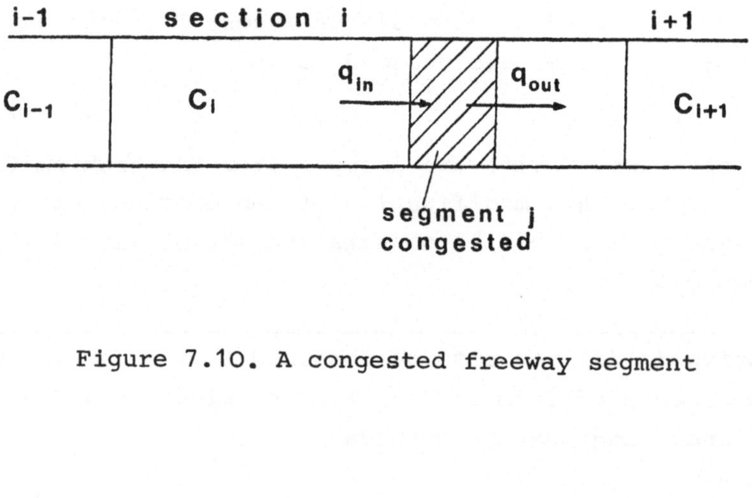

Figure 7.10. A congested freeway segment

with

$$q_{in}(k) = Q\,[C_i(k),\ b_i(k)] \tag{7.42}$$

$$q_{out}(k) = Q\,[\rho_i^j(k),\ b_i(k)]. \tag{7.43}$$

In view of eqns. (7.41)-(7.43), a modification of the continuity equation for the i-th and (i+1)st section becomes then necessary

$$C_i(k+1) = C_i(k) + \frac{T\alpha}{\delta_i}\{Q[C_{i-1}(k),\ b_{i-1}(k)] - Q\,[C_i(k),\ b_i(k)]\} \tag{7.44}$$

$$C_{i+1}(k+1) = C_{i+1}(k) + \frac{T}{\delta_{i+1}}\ \{q_{out}(k) - \alpha . Q\,[C_{i+1}(k),\ b_{i+1}(k)] - (1-\alpha)\ Q\,[C_{i+2}(k),\ b_{i+2}(k)]\}. \tag{7.45}$$

Thus, if a congestion is present in section i, a parameter value p=i should be set in the adaptation layer leading to the solution of a modified optimization problem as described by eqns. (7.41) - (7.45). Absence of congestion should be indicated by p=0. Let us formulate the simplified dynamic optimization problem P10:

> For given predicted trajectories $\underline{d}_N$ and $\underline{\gamma}_N$, parameter $\underline{p}$ and initial condition $\rho_i^j(0)$, $\underline{C}(0)$, $\underline{l}(0)$ find the sequence $[\underline{r}_N^T,\ \underline{b}_N^T]_{(k)}$, k=0,..., K'-1 minimizing the total time spent (7.39) subject to equations (5.2), (5.7), (7.37), (7.40)-(7.45).

The solution of problem P10 will be called the nominal solution.

Remarks:

1) The order of the simplified problem has been considerably reduced. For example, traffic control on a freeway 50 km in length with 10 on-ramps and 10 off-ramps would require a mathematical model of the order 110 in the original formulation. The simplified formulation results in a model of order 20.

2) The computer time and storage space required for the solution of the optimization problem will be furthermore reduced, since longer sample time periods T' > T are used.

3) Occurence of singular subarcs is not excluded in the simplified solution as well. Specific algorithms should therefore be applied /118/.

4) The main properties of the freeway traffic process are still considered in the simplified model. Hence, traffic control obtained by solution of P10 is expected to approximate the solution of P3.

5) As an alternative dynamic optimization problem, the dynamic time-of-day strategies described in section 5.1d) could be utilized in the optimization layer.

a2) Steady-state optimization

Slow disturbances d and A can be assumed constant over time periods of 15-30 minutes /10,68/. This time period is long enough for the traffic flow on short freeways to reach a steady-state. Hence, specification of an optimal steady-state condition can be considered as an objective of the optimization layer, whilst direct control layer is concerned with the transition of the current traffic state to its optimal steady-state point.

The steady-state time-of-day control described in section 5.1.c) can be used for this purpose. Application of steady-state optimal control $\bar{\underline{u}}_N$ is supposed to drive the current process state to the according nominal steady-state point $(\bar{\underline{\rho}}_N, \bar{\underline{v}}_N)$. In case of congestion however, it is possible to achieve the steady-state point, only if appropriately specified values of the maximum admissible volumes $\underline{q}_{max}$ are considered in P1, as will be shown in the discussion of the adaptation layer.

b) <u>The direct control layer</u>

It is the task of the direct control layer to keep the actual values of the state and control variables in the vicinity of the nominal trajectories (or nominal state point), eliminating the effect of slight unexpected disturbances, model inaccuracies or initial deviations. For this purpose, the following quadratic performance functional should be minimized

$$J_Q = \frac{1}{2} \sum_{k=0}^{\infty} [\, \| \underline{x}(k) - \underline{x}_N(k) \|_Q^2 + \| \underline{r}(k) - \underline{r}_N(k) \|_R^2 \,] \qquad (7.46)$$

with $Q = \mathrm{diag}\,(0,1,0.3, \ldots, 1, 0.3)$ and
$R = \mathrm{diag}\,(0.01, \ldots, 0.01)$.

Linearization of the system equations (3.1), (3.2) and formulation of an LQ-Problem as proposed in /119, 120/ cannot drive traffic state back to its nominal state in case of strong disturbances, like incidents.

In other words, the region of attraction $\Lambda \subset R^{N+2n}$ of the closed-loop system around the nominal trajectory or nominal state point does not include all physically meaningful state points (figure 7.5), For $\underline{x} \in \Lambda$, the closed-loop system remains stable (no congestion). If due to a strong disturbance (incident), $\underline{x}$ is driven outside Λ , a congestion will occur.

In spite of this fact, we will minimize the criterion (7.46) subject to the linearized equations and we will guarantee achievability of the nominal state under severe disturbance conditions through appropriate design of the adaptation layer. Solution of the LQ-Problem results to the feedback law

$$\underline{r}(k) = \underline{r}_N(k) - L(\underline{x}_N, \underline{u}_N, A_N, k) \cdot [\underline{x}(k) - \underline{x}_N(k)] \qquad (7.47)$$

where matrix L can be computed from a matrix - difference equation. Calculation of L for some representative situations showed that its rows L_i take similar values around on-ramp segments. Let us denote by $\underline{L}_{i,\rho}$ the vector

$$\underline{L}_{i,\rho} = [L_{i,1}, L_{i,3}, \ldots, L_{i,\ 2n-1}]$$

feeding back the density deviations from the nominal trajectory. Typical values for $L_{i,\rho}$ are given in figure 7.11. Bacause of the overlapping character of the feedback law, decentralized or overlapping control schemes can be applied without any substantial rise of the performance index /113, 114/.

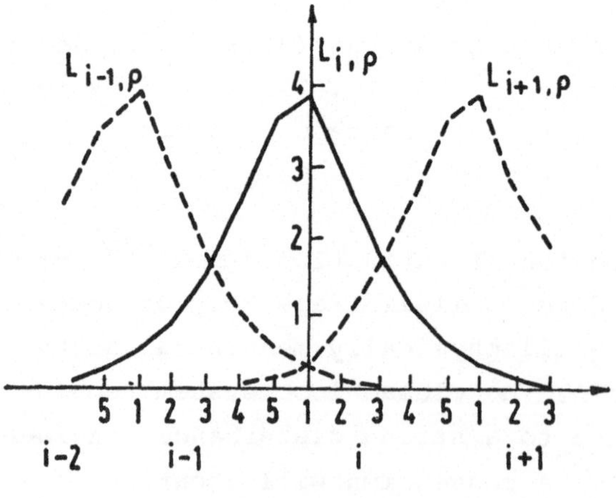

Fig. 7.11-Typical feedback values $L_{i,\rho}$

Of course, in the case of erroneous predictions $\underline{d}_N$, A_N, the nominal trajectories $\underline{x}_N$, $\underline{u}_N$ are inadequate and the action of direct control layer cannot generally lead to an improvement of the results. This is an additional reason why an adaptation layer is introduced.

c) The Adaptation Layer

It is the task of the adaptation layer to specify the values of the predicted disturbances and of the parameter $\underline{p}$ (or $\underline{q}_{max}$) so as to guarantee robustness and stability of the overall system. The information needed for this purpose is provided by local measurement devices (figure 7.12).

c1) Adaptation of slow disturbances

In the problem P10 (or P1) of the optimization layer, $\underline{d}_N$ and A_N are supposed to have known values. Any significant deviation of the current demand and origin-destination values from their predicted values leads to a nonoptimal, possibly inadequate solution. The adaptation layer has to intervene any time such a situation occurs initiating a new optimization run with corrected values of the predicted disturbance trajectories. In order to avoid unnecessarily high switch rates simple pattern classification algorithms are applied /121/.

The region of possible values of each disturbance is divided into several subregions corresponding to specific nominal values.

The question to be answered is: When should a new optimization run be started? If the answer depends simply on the current values $|z_i(k) - z_{i,N}(k)|$, high switch rates could result at the board of two subregions. For this reason, a feature vector $\underline{\eta}$ /121/ is created in the adaptation layer

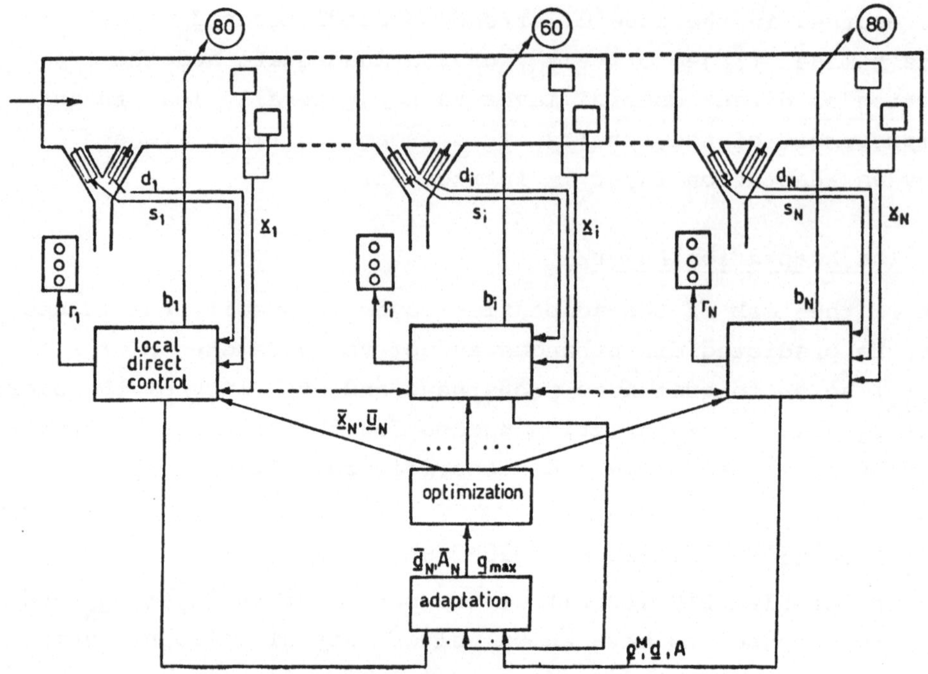

Figure 7.12: A multilayer control system for freeway traffic.

including

- actual deviation $|z_i(k) - z_{i,N}(k)|$
- actual subregion
- time of exceeding the subregion's board
- mean value of the last 2o sampling intervals.

The space of feature vectors Ω is subdivided into classes ω_i corresponding to the disturbance subregions. According to the actual value of the feature vector, a decision can be made whether a class change is necessary or not. In the first case, a new optimization run with a corrected value of the according disturbance is intiated /67/.

c2) Adaptation of the parameter p (or q_{max})

In the case of the dynamic optimization problem P10, parameter p should be set equal to the section number in which a congestion has occured. In order to decide whether a congestion is present or not, simple pattern classification algorithms like the one described above are applied.
In the case of the static optimization problem P1, some maximum volume values $q_{i,max}$ should be specified in the adaptation layer. If no congestion is present, $q_{i,max}$ are set equal to the section's capacity, as indicated in eqn. (5.21). However, in case of congestion, $q_{i,max}$ values should be altered so as to guarantee elimination of congestion. In order to see that, consider a congestion in section i. As mentioned before (see eqn. (7.43)), outflow q_{out} of the section in that case mainly depends upon the congestion density. Elimination of the congestion is possible olny if $q_{in} < q_{out}$. But according to eqn. (7.41), q_{in} is approximately egual to $\bar{q}_i$ which is given by eqn. (2.15). Hence, in order to guarantee elimination of the congestion, we must require

$$\bar{q}_i \leq q_{out} < q_K(\bar{b}_i) \qquad (7.48)$$

which corresponds to a new $q_{i,max}$ setting. The value of $q_{i,max}$ is specified according to the following rule

$$\begin{aligned} &\text{class } \omega_1 : \text{ normal flow } q_{i,max} = q_K(b_i) \\ &\text{class } \omega_2 : \text{ congestion } q_{i,max} = 2000 \qquad (7.49) \\ &\text{class } \omega_3 : \text{ severe congestion } q_{i,max} = 1400. \end{aligned}$$

The underlying idea can be demonstrated on the basis of figure 7.5. The region of attraction of the nominal state point depends on $q_{i,max}$ and $\underline{z}_N$. Hence, any time the current process state $\underline{x}$

is driven outside Λ by some strong disturbance, an appropriate change of $\underline{q}_{max}$ (or p) as described above, can lead to a new nominal state point ($\underline{x}_N'$, $\underline{u}_N'$) and a new region of attraction Λ' including the current state. In that way, the current traffic state can be driven back to the normal traffic region Λ. When this is achieved, a normal value of $\underline{q}_{max}$ (or p) according to eqn. (7.49) is again chosen.

The values of $q_{i,max}$ in eqn. (7.49) have been chosen according to extensive congestion observations /122,123/. The main variable leading to the decision about the actual traffic class of a section is the maximum density

$$\rho_i^M = \max \{\rho_i^1, \rho_i^2, \ldots, \rho_i^{n(i)}\} \; . \tag{7.50}$$

For the selection of the corresponding traffic class simple pattern classification algorithms are used.

d) Summary

The overall control actions are now summarized:

(i) The adaptation layer specifies the predicted trajectories $\underline{d}_N$, A_N and parameter p(or $\underline{q}_{max}$).

(ii) The optimization layer specifies through solution of P1 or P10 the nominal solution.

(iii) After calculation of the feedback matrix L, the direct control actions can be activated.

(iv) In case of a significant deviation of the predicted from the current slow disturbances or in case of a congestion due to an unexpected incident, the adaptation layer proceeds to a new specification of $\underline{d}_N$, A_N and p (or $\underline{q}_{max}$) and initiates a new optimization run.

Figure 7.12 shows the structure of a three layer control system with decentralized direct control layer.

e) <u>Simulation results</u>

In order to study the efficiency of the developed multilayer control structure, the hypothetical freeway of section 6.2.a) will now be considered. Solution of the steady-state time-of-day control problem P1 with the on-ramp demand values given in section 6.2.a) and the O-D-Matrix

$$\underline{A} = \begin{bmatrix} 1 & & & & & \\ 0.95 & 1 & & & 0 & \\ 0.9 & 0.95 & 1 & & & \\ 0.85 & 0.9 & 0.95 & 1 & & \\ 0.8 & 0.86 & 0.9 & 0.95 & 1 & \\ 0.75 & 0.8 & 0.85 & 0.9 & 0.85 & 1 \end{bmatrix}$$

results to the nominal values $\underline{\rho}_N$, $\underline{v}_N$, $\underline{r}_N$ given in section 6.2.a).

e1) <u>Transition to the nominal state point</u>

Consider the initial condition

$$\rho_1^j = 40; \; \rho_i^j = 60; \quad i = 2,3,4,6; \; \rho_5^j = 75; \quad j = 1,\dots,5$$

$$v_i^j = 80; \; v_i^j = 70; \quad i = 2,3,4,6; \; v_5^j = 50; \quad j = 1,\dots,5$$

which means that the road if fairly loaded and traffic density is slightly overcritical in the 5th section. Figure 7.13 shows the results for application of

a) no control action
b) steady-state time-of-day control
c) nonlinear optimal control according to section 6.2.a)
d) multilayer control.

Figure 7.13.a) shows that traffic flow becomes unstable in the uncontrolled case. A severe congestion is built in the 5th section. Besides, a recurrent congestion is built in the

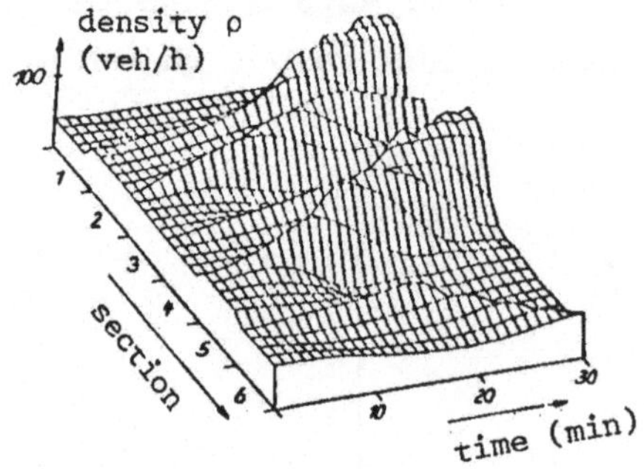

a) without control

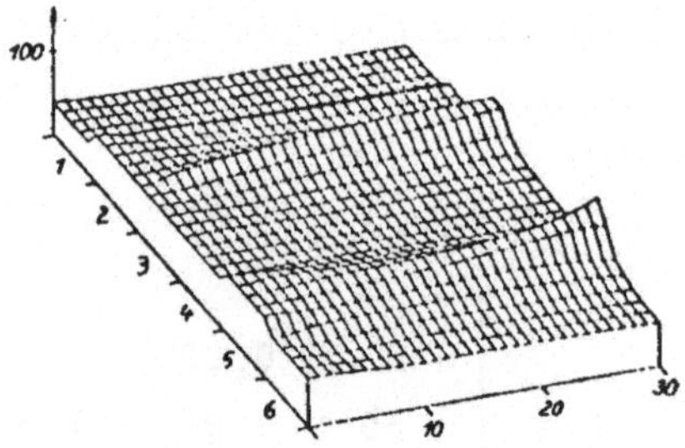

b) time-of-day control

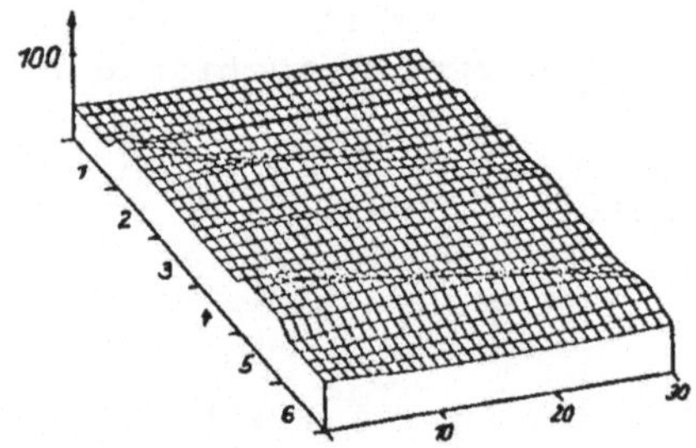

c) optimal control

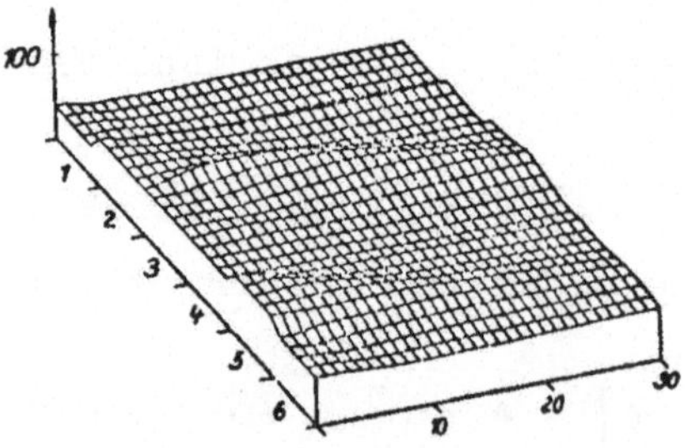

d) multilayer control

Figure 7.13: Evolution of traffic density for the cases a) - d)

3rd section due to high on-ramp demands. A much better situation can be observed in the case of time-of-day control (figure 7.13 b)). A recurrent congestion is prevented and a non recurrent one is built up very slowly. This indicates that the initial state has been in the region B-A of figure 7.6. Indeed, direct control layer succeeds to drive the initial overcritical

traffic state back to its nominal condition without intervention of the adaptation layer, as shown in figure 7.13d). These results are similar to the ones of figure 7.13c), which have been obtained with a much higher computational effort as has already been discussed in section 6.1 and 6.3.
Figure 7.14 shows the corresponding on-ramp trajectories of the 6-th section. A slight reduction of the on-ramp volumes from their nominal values in the cases c) and d) leads to an elimination of the overcritical density.

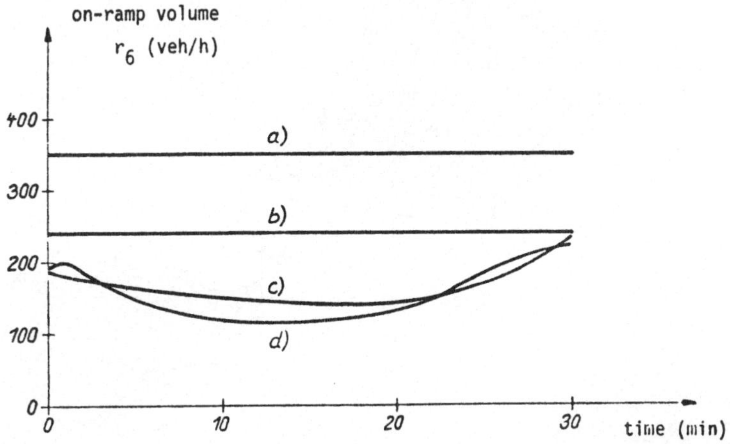

Figure 7.14: On-ramp trajectories for the cases a - d).

e2) Elimination of a congestion

Let us now consider the congested initial condition of section 6.2.a) for the following cases:

a,b) like in the previous simulation
c) Two-layer system (without adaptation)
d) Non linear optimal control of section 6.2.a)
e) Three-layer control.

Results are shown in figure 7.15.
In case a), congestion caused by an accident destabilizes traffic flow. In addition, freeway capacity is exceeded and

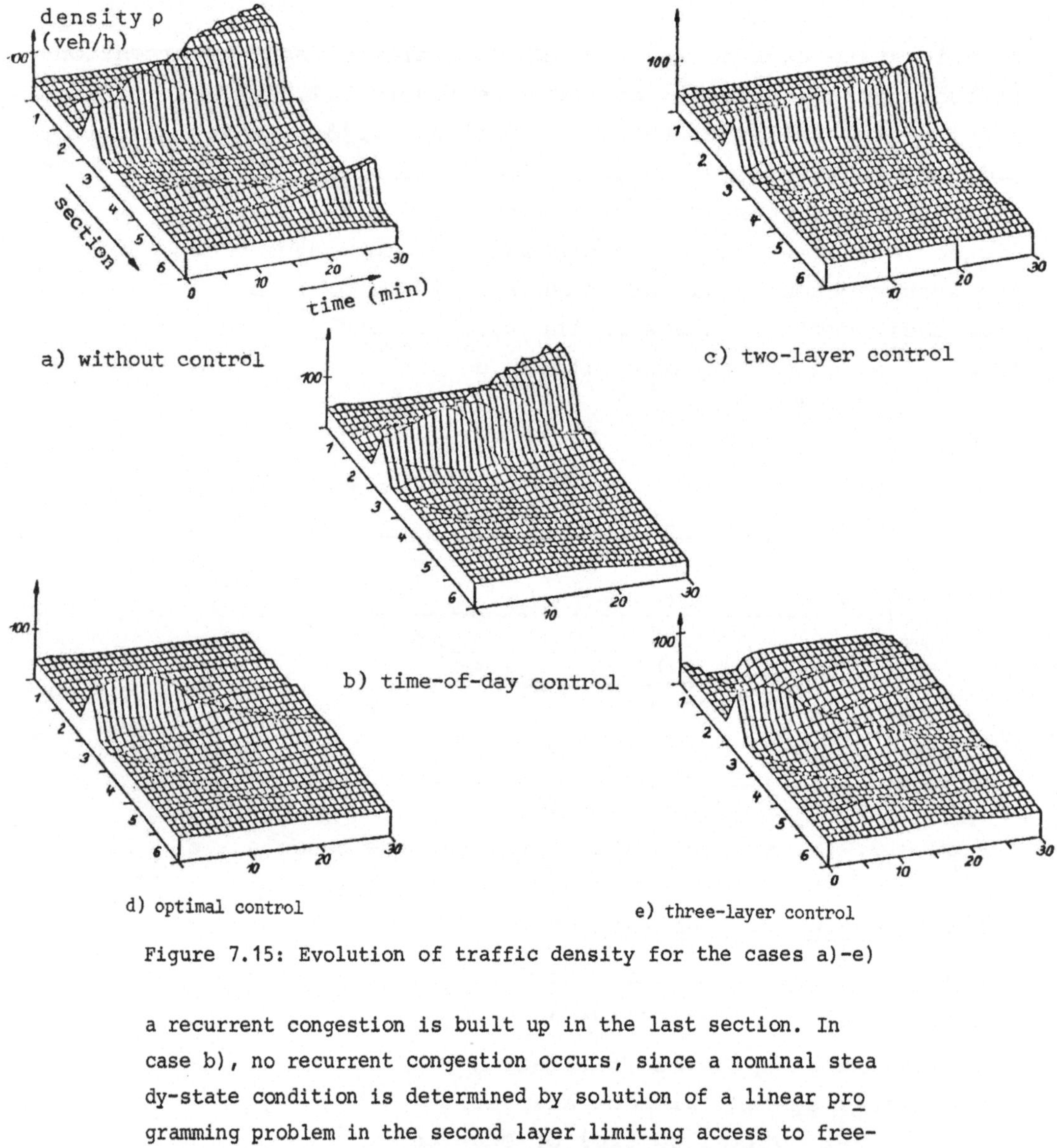

a) without control

b) time-of-day control

c) two-layer control

d) optimal control

e) three-layer control

Figure 7.15: Evolution of traffic density for the cases a)-e)

a recurrent congestion is built up in the last section. In case b), no recurrent congestion occurs, since a nominal steady-state condition is determined by solution of a linear programming problem in the second layer limiting access to freeway capacity. However no elimination of the nonreccurent congestion can be succeded, since initial condition lies in the region C-B of Figure 7.6. For the same reason, congestion can not be eliminated even by application of the feedback law of the direct control layer as shown in figure 7.15.c).

In case d), which is identical to the one of figure 6.5, non-recurrent congestion is successfully eliminated.
In case e), actual traffic conditions are classified to *severe congestion in the 3rd section* and a modified $q_{3,max}$ value is selected in the adaptation layer according to eqn. (7.49).
An optimization run is then started taking into account the reduced freeway capacity due to the congestion. Solution of the new optimization problem leads to a modified steady-state set point for the direct control layer:

$$\bar{\underline{r}}'_N = [1339 \quad 100 \quad 100 \quad 350 \quad 350 \quad 350]^T$$

$$\bar{\underline{b}}'_N = [0.9 \quad 0.8 \quad 0.8 \quad 1 \quad 1 \quad 1]^T$$

$$\bar{\underline{c}}'_N = [12 \quad 12 \quad 12 \quad 19 \quad 19 \quad 25]^T$$

$$\bar{\underline{v}}'_N = [113 \quad 113 \quad 113 \quad 105 \quad 105 \quad 105]^T.$$

After about 10 minutes, traffic state enters the stability region of the class *normal traffic* and the optimal steady-state condition can be specified by means of a new optimization run with normal values for all $q_{i,max}$. Congestion has been eliminated.
The corresponding trajectories of the on-ramp volumes for the first three sections are shown in Figures 7.16 - 7.18.
In case c), because of the decentralized nature of the feedback law, only on-ramp volume r_3 is reduced. On-ramp volume r_2 is reduced only after the congestion has reached section 2. Thus, elimination of congestion is not achieved.
In the case d), congestion is eliminated in a smooth way by reducing on-ramp volumes 2 and 3 as much as necessary. On the other hand, the multilayer structure (case e)) also eliminates the congestion by reduction of on-ramp volumes 1-3 but in a more "rude" way because of its switching strategy and the

classification of the actual state into one of three classes. Although the multilayer system is far not as "fine" as the one of case d), it will be much more robust in a practical application where estimates of the initial condition might be inaccurate.

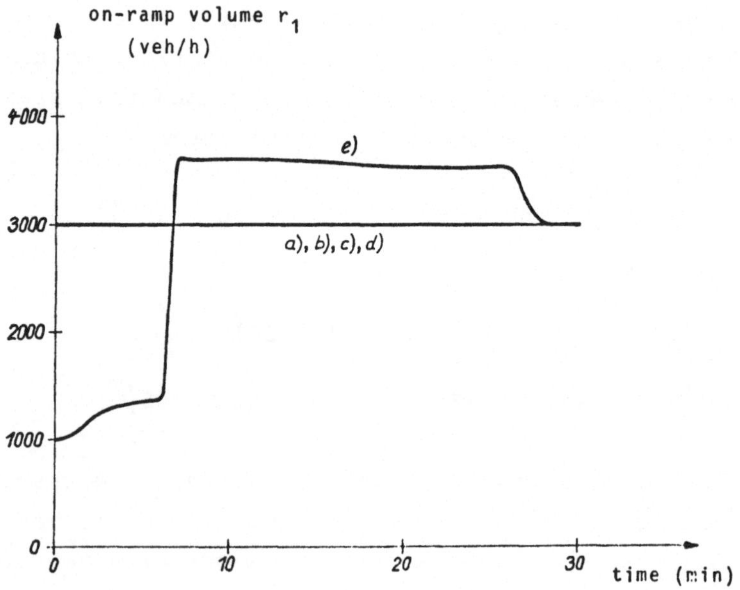

Figure 7.16: On-ramp volume trajectories for section 1

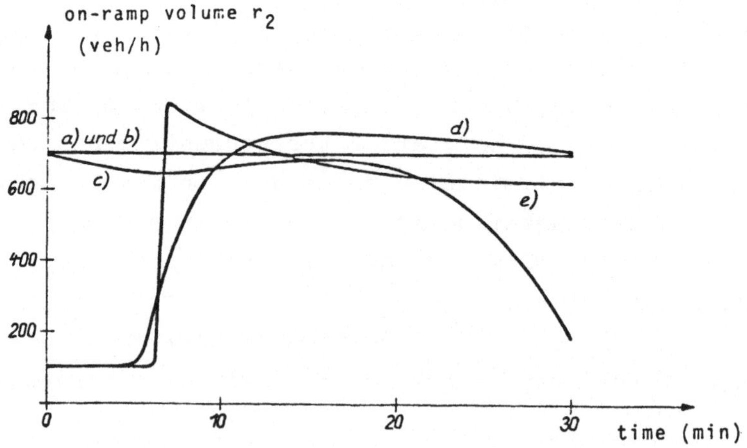

Figure 7.17: On-ramp volume trajectories for section 2

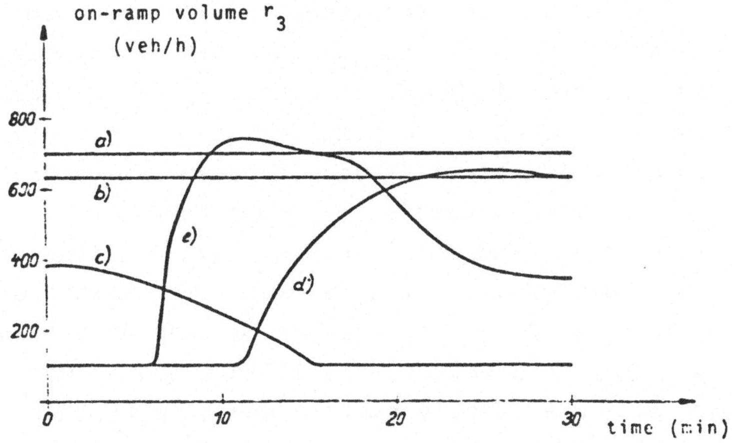

Figure 7.18: On-ramp volume trajectories for section 3.

7.3 Implementation aspects

The state estimation algorithms presented in section 4 provide estimates for the traffic variables between two subsequent detector locations. The estimates are derived on the basis of the measurements coming from the two detectors, so that a natural decentralization of the filtering algorithm for a long freeway can be achieved. Hence, local data-processing, local state estimation and local direct control algorithms can be implemented in a single microcomputer installed at specific freeway locations.

A medium size minicomputer evolves the central tasks of adaptation and optimization as well as input and output tasks like visualization of the current process conditions, storage of process data needed for statistic analysis etc. Since only a small part of the measured data are fed back to the adaptation layer, rather low communication data rates between the central computer and the local microcomputers are expected. Inexpensive telephone channels can be used for the data transfer /124/.

For the particular estimation algorithms of /44/, each local microcomputer station is configured as an 8-bit microprocessor with an attached arithmetic processing unit (APU). The estimation algorithm and the direct control layer for one freeway section require approximately 1.5kByte of computer storage space for the application programs. The computation time needed for evaluation of the same tasks at each sampling period is in the order of 1s, which is much lower than the sampling time interval of T_s=15s. Because of the low storage and computation time requirements, surveillance and direct control of more than one freeway section can be performed in the same local microcomputer station.

8. CONCLUSIONS AND FUTURE DEVELOPMENTS

There has been a substantial amount of research work devoted to urban traffic control systems in the last two decades. With respect to freeway control systems a number of powerful algorithms for modelling, estimation and control have been developed.
The macroscopic freeway traffic model by Payne /51/, which has been referred throughout this monograph model D, seems to provide a reasonable tool for simulation, estimation and control of freeway systems.
In fact, Payne's model is used in almost all recent freeway traffic control algorithms /27,29,44,45,57,59,67,87,98,114, 125,126,127/. Possible improvements of the model's performance are discussed by Paune in /32/.

Efficient estimation algorithms have been developed by various researchers as described in section 4. They provide accurate estimates of traffic variables between two detection locations. Some of these algorithms are capable of estimating traffic state even for a relatively long detector distance leading to a substantial reduction of the implementation effort. A special research topic of freeway traffic estimation is the important problem of incident detection. First results obtained are very encouraging but additional research work seems to be necessary for the development of practical incident detection systems.

With respect to control algorithms, the multilayer framework seems to provide the most important practicable tool for freeway traffic control. Although the concrete optimization problems treated in each control layer may differ according to particular application conditions, the main idea of:

- solving a simplified optimization problem subject to the overall system (optimization layer) on the basis of
- updated predictions of traffic conditions (adaptation layer) and performing
- feeback control laws to account for model simplifications and external disturbances (direct control layer)

is implicitly or explicitly applied to the most proposed modern traffic control systems /9,10,11,62,65,67,87,119,126/. On the other hand, hierarchical optimization algorithms seem to provide a useful tool for evaluation of particular traffic problems.

In distinction to the freeway traffic control problem, there don't seem to be any generally recognized models and control algorithms for road networks or corridor systems. The pioneering research work by Gazis (section 5.2) for a single intersection included application of the Pontryagin's Maximum Principle. Its extension to traffic networks is from a theoretical point of view straightforward but enormeous numerical difficulties arise if the dimension of the problem exceeds certain limits. On the other hand, results provided by Singh and Tamura (section 6.2) are of a preliminary nature and cannot be directly applied to practical situations. Thus, there is an apparent need for powerful algorithms for network traffic control, applicable to cases of several road crosses and taking into account practical requirements like computation time, real-time treatment, implementation effort etc. New approaches like the one of Baras and Levins (section 2.3) might be useful in that context.

Appendix 1: Kalman Filters

A1.1 The Kalman-Filter for linear systems

a) Problem Formulation

Consider the time-discrete state equations

$$\underline{x}(k+1) = A(k).\ \underline{x}(k) + B(k)\ \underline{u}(k) + D(k)\ \underline{\gamma}(k) \tag{A1.1}$$

and the output vector equations

$$\underline{y}(k) = C(k)\ \underline{x}(k) + \underline{\zeta}(k) \tag{A1.2}$$

where the system noise $\underline{\gamma}(k)$ and measurement noise $\underline{\zeta}(k)$ are white vector gaussian sequences with zero means and known covariance matrices, i.e.

$$E\ \{\underline{\gamma}(k)\} = \underline{O} \quad \text{and} \quad E\ \{\underline{\gamma}(k_i)\underline{\gamma}^T(k_j)\} = Q(k_i).\delta_{ij} \tag{A1.3}$$

$$E\ \{\underline{\zeta}(k)\} = \underline{O} \quad \text{and} \quad E\ \{\underline{\zeta}(k_i)\underline{\zeta}^T(k_j)\} = R(k_i).\delta_{ij} \tag{A1.4}$$

with

$$\delta_{ij} = \begin{cases} 0 & \text{for} \quad i \neq j \\ 1 & \text{for} \quad i = j \end{cases} \tag{A1.5}$$

and $Q(k_i) \geq 0$ and $R(k_i) > 0$.
The initial condition $\underline{x}(k_o)$ of the system equation is considered as a gaussian random variable with known mean value

$$\underline{x}(k_o) = E\ \{\underline{x}(k_o)\} \tag{A1.6}$$

and covariance matrix

$$\underline{P}(k_o) = E\{[\underline{x}(k_o) - \hat{\underline{x}}(k_o)] [\underline{x}(k_o) - \hat{\underline{x}}(k_o)]^T\} \quad (A1.7)$$

$\underline{x}(k_o)$ is supposed to be uncorrelated with $\underline{\gamma}(k_i)$ and $\underline{\zeta}(k_i)$, i.e.

$$E\{\underline{\gamma}(k_i)\ \underline{x}(k_o)^T\} = E\{\underline{\zeta}(k_i)\ \underline{x}(k_o)^T\} = 0\ . \quad (A1.8)$$

$\underline{\gamma}(k_i)$ and $\underline{\zeta}(k_i)$ can be correlated

$$E\{\underline{\gamma}(k_i)\ \underline{\zeta}(k_i)^T\} = M(k_i).\delta_{ij}\ . \quad (A1.9)$$

On the basis of known measurments $\underline{y}(k)$ and $\underline{u}(k)$ at time k, we wish to reconstruct the state variable so as to minimize the mean squares error

$$E\{\tilde{\underline{x}}(k)\ \tilde{\underline{x}}(k)^T\} \longrightarrow \min \quad (A1.10)$$

where

$$\tilde{\underline{x}}(k) = \underline{x}(k) - \hat{\underline{x}}(k) \quad (A1.11)$$

is the estimation error.

b) Solution /52/

The estimated state can be calculated by on-line treatment of the following equations

$$\hat{\underline{x}}(k+1) = A(k)\ \hat{\underline{x}}(k) + B(k).\underline{u}(k) + K(k)\ [\underline{y}(k) - C(k)\ \hat{\underline{x}}(k)] \quad (A1.12)$$

with

$$K(k) = [A(k)P(k)C(k)^T + B(k)\ M(k)][C(k)P(k)C(k)^T + R(k)]^{-1} \quad (A1.13)$$

where P(k) is the covariance matrix of the estimation error

$$P(k) = E\{\underline{\tilde{x}}(k)\ \underline{\tilde{x}}(k)^T\} \tag{A1.14}$$

and can be calculated by the matrix difference equation

$$P(k+1) = [A(k)-K(k)C(k)]\ P(k)A(k)^T+B(k)Q(k)B(k)^T$$
$$-K(k)M(k)^T B(k)^T . \tag{A1.15}$$

The initial condition of the difference equation (A1.12) and (A1.15) are given by eqns. (A1.6) and (A1.7) correspondingly.

A1.2 An extended Kalman-Filter for nonlinear systems

Now consider nonlinear system and measurment equations

$$\underline{x}(k+1) = \underline{f}[\underline{x}(k), \underline{u}(k), \underline{\gamma}(k)] \tag{A1.16}$$

$$\underline{y}(k) = \underline{g}[\underline{x}(k)] + \underline{\zeta}(k). \tag{A1.17}$$

I is known that no finite dimensional optimal filter exists for this problem /53/. A plausible suboptimal solution can be specified, if the system, input and measurment matrices needed for equations (A1.13), (A1.15) are on-line calculated at each time instant through linearization of the nonlinear equations (A1.16), (A1.17) around the current estimated state $\underline{\hat{x}}(k)$ /54/, i.e.

$$A = \frac{\partial \underline{f}}{\partial \underline{x}} \Bigg|_{\substack{\underline{x} = \underline{\hat{x}}(k) \\ \underline{u} = \underline{u}(k) \\ \underline{\gamma} = \underline{0}}} \qquad B = \frac{\partial \underline{f}}{\partial \underline{u}} \Bigg|_{\substack{\underline{x} = \underline{\hat{x}}(k) \\ \underline{u} = \underline{u}(k) \\ \underline{\gamma} = \underline{0}}} \tag{A1.18}$$

$$C = \frac{\partial \underline{g}}{\partial \underline{x}} \bigg|_{\underline{x} = \hat{\underline{x}}(k)} . \tag{A1.19}$$

On the other hand, equation (A1.12) is now written in terms of the nonlinear model, i.e.

$$\hat{\underline{x}}(k+1) = \underline{f}[\hat{\underline{x}}(k), \underline{u}(k), \underline{O}] + K(k) [\underline{y}(k) - \underline{g}[\hat{\underline{x}}(k)]]. \tag{A1.20}$$

It should be noted at this point that because of the linearization performed with equations (A1.18) and (A1.19), the computational effort needed for the extended Kalman Filter can be much higher compared to the standard linear case. Besides, convergence of the estimated state $\hat{\underline{x}} \rightarrow \underline{x}$ cannot be generally guaranteed.

Appendix 2: The Pontryagin's Maximum Principle

A2.1 Continuous-time formulation

a) Problem formulation

Consider the nonlinear dynamic system

$$\dot{\underline{x}} = \underline{f}(\underline{x},\underline{u},t) \quad ; \quad \underline{x}(o) = \underline{x} \tag{A2.1}$$

with dim($\underline{x}$) = n. The admissible state and control region is defined by

$$\underline{h}(\underline{x},\underline{u},t) \geq \underline{O} \tag{A2.2}$$

where $\underline{h}$ is assumed to fulfill some constraint conditions given in /79/. The state at the final time point should fulfill

$$\underline{N}[\underline{x}(t_e),\ t_e] = \underline{O} \tag{A2.3}$$

where dim$\underline{N} \leq$ dim$\underline{x}$. The problem consists in finding a control input trajectory minimizing the performance functional

$$F = \theta[\underline{x}(t_e),t_e] + \int_{0}^{t_e} \phi(\underline{x},\underline{u},t)dt \tag{A2.4}$$

subject to eqns. (A2.1) - (A2.3).

b) Solution /79/

Define the Hamiltonian

$$H = \phi(\underline{x},\underline{u},t) + \underline{\lambda}^T\ \underline{f}(\underline{x},\underline{u},t) \tag{A2.5}$$

with $\underline{\lambda}$ the n-dimensional continuous, non vanishing vector of costate variables. An optimal input trajectory $\underline{u}^*(t)$ must fulfill the following necessary conditions for optimality:

$$\underline{\dot{x}} = \frac{\partial H}{\partial \underline{\lambda}} = \underline{f}(\underline{x},\underline{u},t) \tag{A2.6}$$

$$\underline{\dot{\lambda}} = \frac{-\partial H}{\partial \underline{x}} - \frac{\partial \underline{h}}{\partial \underline{x}}^T \underline{\mu} \tag{A2.7}$$

$$\frac{\partial H}{\partial \underline{u}} + \frac{\partial \underline{h}}{\partial \underline{u}}^T \underline{\mu} = \underline{O} \tag{A2.8}$$

$$\underline{\mu}^T \underline{h} = 0, \quad \underline{\mu} \leqslant \underline{O}, \quad \underline{h}(\underline{x},\underline{u},t) \geqslant \underline{O} \tag{A2.9}$$

$$H(t,\underline{x}^*,\underline{u},\underline{\lambda}) \geqslant H(t,\underline{x}^*,\underline{u}^*,\underline{\lambda}) \tag{A2.10}$$

and the boundary conditions

$$\underline{x}(0) = \underline{x}_0 \tag{A2.11}$$

$$\underline{N}[\underline{x}(t_e),t_e] = \underline{O} \tag{A2.12}$$

$$\frac{\partial \theta[x(t_e),t_e]}{\partial \underline{x}} - \underline{\lambda}(t_e) + \frac{\partial N[\underline{x}(t_e),t_e]^T}{\partial \underline{x}} \underline{\nu} = \underline{O} \tag{A2.13}$$

$$\left[H(t_e) + \frac{\partial \theta[\underline{x}(t_e),t_e]}{\partial t}\right] \delta t_e = 0 \tag{A2.14}$$

with $\underline{\mu}(t)$ and $\underline{\nu}$ some additional multipliers.
If the constraints (A2.2) have the form

$$B^i(t,\underline{x}) \leqslant u^i \leqslant A^i(t,\underline{x}) \tag{A2.15}$$

where $A^i > B^i$, then conditions (A2.8), (A2.9) imply that

$$\frac{\partial H}{\partial u^i} \begin{cases} \geq 0 & \text{if} \quad u^{*i} = B^i \\ = 0 & \text{if} \quad B^i < u^{*i} < A^i \\ \leq 0 & \text{if} \quad u^{*i} = A^i \end{cases} \tag{A2.16}$$

so that eqn. (A2.16) can be used in place of eqns. (A2.8), (A2.9).

If $\underline{h}$ is independent of $\underline{x}$, then eqn. (A2.7) becomes

$$\dot{\underline{x}} = - \frac{\partial H}{\partial \underline{x}} \tag{A2.17}$$

and $\underline{u}^*$ can be specified by eqn. (A2.10) so that $\underline{\mu}$ has no influence upon the results and hence eqns. (A2.8), (A2.9) should not necessarily be considered. The latter case has been the one considered by Pontryagin /80/.

A2.2 Discrete-time formulation

α) Problem formulation

Let us now consider a particular discrete-time analogon of A2.1. The state equations of the considered system are given by

$$\underline{x}(k+1) = \underline{f}[\underline{x}(k), \underline{u}(k), k] \; ; \; \underline{x}(0) = \underline{x}_0 \tag{A2.18}$$

with inequality constraints

$$\underline{h}[\underline{x}(k), \underline{u}(k), k] \geq \underline{0} \tag{A2.19}$$

and performance functional to be minimized

$$J = \theta[\underline{x}(K)] + \sum_{k=0}^{K-1} \phi[\underline{x}(k),\underline{u}(k),k]. \tag{A2.20}$$

The end time K is considered fixed, and the end time state $\underline{x}(T)$ is free.

b) Solution /81/

Define the Hamiltonian

$$H = \phi[\underline{x}(k),\underline{u}(k),k] + \underline{\lambda}(k+1)^T \underline{f}[\underline{x}(k),\underline{u}(k),k]. \tag{A2.21}$$

An optimal input trajectony $\underline{u}^*(k), k = 0,\ldots,K-1$, should fulfill the following necessary conditions

$$\underline{x}(k+1) = \frac{\partial H}{\partial \underline{\lambda}(k+1)} = \underline{f}[\underline{x}(k),\underline{u}(k),k] \tag{A2.22}$$

$$\underline{\lambda}(k) = \frac{\partial H}{\partial \underline{x}(k)} + \frac{\partial \underline{h}}{\partial \underline{x}(k)}^T \underline{\mu}(k) \tag{A2.23}$$

$$\frac{\partial H}{\partial \underline{u}(k)} + \frac{\partial \underline{h}^T}{\partial \underline{u}(k)} \underline{\mu}(k) = \underline{0} \tag{A2.24}$$

$$\underline{\mu}(k)^T \underline{h} = 0 \quad , \quad \underline{\mu} \leq \underline{0} \quad , \quad \underline{h}[\underline{x}(k),\underline{u}(k),k] \geq \underline{0} \tag{A2.25}$$

and the boundary conditions

$$\underline{x}(0) = \underline{x}_0 \tag{A2.26}$$

$$\underline{\lambda}(K) = \partial\theta[\underline{x}(K)]/\partial\underline{x}. \tag{A2.27}$$

The time-discrete analogon of eqn. (A2.10) does generally not represent a necessary condition for optimality /81,82/. Simplifications indicated by eqns. (A2.16), (A2.17) are also valid for the time-discrete case.

REFERENCES

/1/ Cannon, R.H., Jr.: Transportation, Automation and Societal Structure. Proc. IEEE 61 (1973), pp. 518-525.

/2/ Evans, J.B.: Microprocessors: A discussion in the context of Traffic Control. Traffic Eng. & Control 19 (1978), pp. 112-115.

/3/ Strobel, H.: Transportation, Automation and the quality of urban living. Intern. Inst. for Appl. Systems Analysis, Res. Rep. 75-34, 1975, 63 p.

/4/ Thedeen, T.: On the aim and future of traffic theoretical research. Transp. Res. 10 (1976), p. 379.

/5/ Obermaier, A.: The "green wave" rolls on.IEEE Spectrum, March 1977, pp. 75-78.

/6/ Torrero, E.A.: Unjamming traffic congestion. IEEE Spectrum, November 1977, pp. 77-79.

/7/ Wardrop, J.G.: Some Theoretical Aspects of Road Traffic Research. Proc. Inst. Civil Engrs., Part II, vol. 1 (1952), pp. 325-362.

/8/ Gazis, D.C., Edie, L.C.: Traffic Flow Theory. Proc. IEEE 56 (1968), pp. 458-471.

/9/ Tabak, D.: Application of Modern Control and Optimization Techniques to Transportation Systems. In "Control and Dynamic Systems" 10 (1973), C.T. Leondes, Ed., Academic Press, pp. 345-434.

/10/ Isaksen, L., Payne, H.J.: Freeway Traffic Surveillance and Control. Proc. IEEE 61 (1973), pp. 526-536.

/11/ May, A.D.: A proposed Dynamic Freeway Control System Hierarchy. Proc. IFAC/IFIP/IFORS 3rd Intern. Symp. on Transportation Systems, Columbus, 1976, pp. 1-12.

/12/ Saxton, L., Schenck, C.: Diversion and Corridor Control Systems in Western Europe. In "World Survey on Current Research and Development on Roads and Road Transport", Intern. Road Federation, Washington, D.C., 1977, pp. 692-723.

/13/ Papageorgiou, M., Schmidt, G.: Steuerungs-und Regelungsaufgaben bei der Beein flussung des Verkehrsablaufs auf Schnellstrassen. Regelungstechnik 26 (1978), pp. 282-291.

/14/ Papageorgiou, M., Posch, B., Schmidt, G.: Comparison of macroscopic Mode ls for Control of Freeway Traffic. Transp. Research, to appear.

/15/ Reuschel, A.: Fahrzeugbewegungen in der Kolonne. Zeitschrift d. Oesterr. Ing. u. Arch. Vereins 95 (1950), pp. 59-62, 73-77.

/16/ Pipes, L.A.: An Operational Analysis of Traffic Dynamics. Jour. Appied Physics 24 (1953), pp. 274-281.

/17/ Gazis, D.C., Herman, R., Potts, R.: Car-following theory of steady-state traffic flow. Oper. Res. 7 (1959), pp. 499-595.

/18/ Hsu, .Y.S., Munjal, P.K.: Freeway digital simulation models. Transp. Res. Rec. 509 (1974), pp. 29-41.

/19/ Greenshields, B.D.: A Study in Highway Capacity. Highway Res. Board Proc. 14 (1934), p. 468.

/20/ May, A.D., Keller, H.E.M.: Non-Integer Car-Following Models. Highway Res. Rec. 199 (1967), pp. 19-31.

/21/ Swik, R.: Beitrage zur Regelung der Langsbewegung von Kraftfahrzeugen mit digitalen Mitteln. Dissertation, Technische Universitat Munchen, 1977.

/22/ Posch, B.: Ein Regelkonzept fur die automatische Zusammenfuhrung zweier Fahrzeugschlangen. Dissertation, Technische Universität München, 1982.

/23/ Lighthill, M.J., Whitham, G.B.: On Kinematic Waves II. A Theory of Traffic Flow on Long Crowded Roads. Proc. of the Royal Society of London, Series A 229 (1955), pp. 317-345.

/24/ Richards, P.I.: Shock waves on the Highway. Oper. Res. (1956), pp. 42-51.

/25/ Berger, C.R., Shaw, L.: Discrete-Event Simulation of Freeway Traffic. Simulation Council Proc., Series 7 (1977), pp. 85-93.

/26/ Jeuken, H.: Ein Verfahren zur Beurteilung von Verkehrssituationen auf zweibahnigen Strassen.Dissertation, RWTechnische Hochschule Aachen, 1977.

/27/ Greenlee, T.L., Payne, H.J.: Freeway Ramp Metering Strategies for Responding to Incidents. Proc. of the 1977 IEEE Conf. on Decision and Control, pp. 987-992.

/28/ Lighthill, M.J., Whitham, G.B.: On kinematic waves I. Flood movement on long rivers. Proc. of the Royal Society of London, Series A 229 (1955), pp. 281-316.

/29/ Cremer, M.,Papageorgiou,M.: Parameter identification for a Traffic Flow Model. Automatica 17 (1981), pp. 837-843.

/30/ Papageorgiou, M.: A new approach to time-of-day control based on a dynamic freeway traffic model. Transp. Res. 14B (1980), pp. 349-360.

/31/ Payne, H.J.: Models of Freeway Traffic and Control. Simulation Council Proc. 1 (1971), pp. 51-61.

/32/ Payne, H.J.: A critical review of a macroscopic freeway model. In "Engineering Foundation Conference on Research Directions in Computer Control of Urban Traffic", Levine, W.S.,Lieberman, E., Fearnsides, J.J., Eds., pp. 251-265.

/33/ Gazis, D.C., Potts, R.B.: The oversaturated intersection. Proc. 2nd Intern. Symp. on Traffic Theory, London, 1963, pp. 221-237.

/34/ Gazis, D.C.: Optimum Control of a System of Oversaturated intersections. Operation Res. 12 (1964), pp. 815-831.

/35/ Ross, P., Gibson, D.: Survey of models for simulating traffic. Simulation Councils Proc. Series 7 (1977), pp. 39-48.

/36/ Baras, J.S., Levine, W.S., Lin, T.L.: Discrete-time Point Processes in Urban Traffic Queue Estimation. IEEE Trans. on Autom. Control Ac-24 (1979), pp. 12-27.

/37/ Baras, J.S., Levine, W.S.: Some results on computer control of urban traffic. Preprints of the 2nd IFAC Symp. on Large Scale Systems Theory and Applications, Toulouse, 24-26 June, 1980, pp. 449-457.

/38/ Baras, J.B., Dorsey, A.J., Levine, W.S.: Estimation of Traffic Platoon Structure from Headway Statistics. IEEE Trans. on Autom. Control AC-24 (1979), pp. 553-559.

/39/ Cremer, M., Papageorgiou, M.: Parameter Identification for a Traffic Flow Model. Preprints of the 5th IFAC Symp. on Identification and System Parameter Estimation, Darmstadt, Sept. 1979, pp. 771-778.

/40/ Leutzbach,W.: Einführung in die Theorie des Verkehrsflusses. Springer-Verlag, 1972.

/41/ Aström, K.J., Eykhoff, P.: System Identification-a Survey. Automatica 7 (1971), pp. 123-162.

/42/ Sage, A.P.: Optimum System Control. Prencice-Hall, 1968.

/43/ Box, M.J.: A new method of constrained optimization and a comparison with other methods. Computer Journal 8 (1965), pp. 42-52.

/44/ Cremer, M., Papageorgiou, M., Schmidt, G.: Einsatz regelungstechnischer Mittel zur Verbesserung des Verkehrsablaufs auf Schnellstrassen. Forschung Strassenbau und Strassenverkehrstechnik 307 (1980), pp. 1-44.

/45/ Papageorgiou, M., Heilmann, M.: Ein Programmsystem zur Simulation des Verkehrsablaufs auf Schnellstrassen, Strassenverkehrstechnik 26 (1982), pp. 142-147.

/46/ Gazis, D.C., Knapp, C.H.: On-line Estimation of Traffic Densities from Time-Series of Flow and Speed Data, Transp. Science 5 (1971), pp. 283-301.

/47/ Szeto, M.W., Gazis, D.C.: Application of Kalman Filtering to the Surveillance and Control of Traffic systems, IBM Research Report RC 3690, 1972, pp. 419-439.

/48/ Nahi, N.E., Trivedi, A.N.: Recursive Estimation of Traffic Variables: Section Density and Average Speed. Transp. Science 7 (1973), pp. 269-286.

/49/ Nahi, N.E.: Freeway-Traffic Data Processing. Proc. IEEE 61 (1973), pp. 537-541.

/50/ Ghosh, D., Knapp, C.H.: Estimation of Traffic Variables Using a Linear Model of Traffic Flow. Transp. Research 12 (1978), pp. 395-402.

/51/ Cremer, M.: Der Verkehrsfluss auf Schnellstrassen. Springer-Verlag, 1979.

/52/ Kalman, R.E., Bucy, R.S.: New Results in Linear Filtering and Prediction Theory. Trans. ASME Series D 83 (1961), pp. 95-108.

/53/ Kushner, H.J.: Dynamical Equations for Optimal Nonlinear Filtering. J. Diff. Equations 3 (1967), pp. 179-190.

/54/ Jazwinsky, A.H.: Stochastic Processes and Filtering Theory. Academic Press, 1970.

/55/ Payne, H.J.: Calibration of Freeway Incident Detection Algorithms. Proc. 3rd IFAC/IFIP/IFORS Symp. on Control in Transp. Systems, Columbus, Aug. 1976, pp. 29-37.

/56/ Levin, M., Krause, G.M.: A probabilistic approach to incident detection on urban freeways. Traffic Engng. & Control, March 1979, pp. 107-109.

/57/ Willsky, A.S., Houpt, P.K., Gershwin, S.B., Kurkijan, A.L., Greene, C.S., Chow, E.V.: Detection of Incidents on Freeways. Proc. 1978 IEEE Conf. on Decision and Control, San Diego, pp. 1037-1041.

/58/ Böttger, R.: Ein Verfahren zur messtechnischen Feststellung von Verkehrsstörungen auf Fernstrassen und Autobahnen. Strassenverkehrstechnik 23(1979), pp. 173-179.

/59/ Cremer, M.: Incident Detection on Freeways by Filtering Techniques. Preprints of the 8th IFAC World Congress, Kyoto, Aug. 1981, pp. XVII-96-101.

/60/ May, A.D., Jr.: Experimentation with manual and automatic ramp control. In "Traffic Control Theory and Instrumentation", T.R. Horton, E.D., Plenum Press, 1965, pp. 157-208.

/61/ Gervais, E.F.: Optimization of Freeway Traffic by Ramp Control. Highway Res. Rec. 59 (1964), pp. 104-118.

/62/ Drew, D.R., Brewer, K.A., Buhr, J.H., Whitson, R.H.: Multilevel Approach to the Design of a Freeway Control System. Highway Res. Rec. 279 (1969), pp. 40-55.

/63/ Ranabauer, A., Abel, E.: Die erste ferngesteuerte Verkehrszeichen - und Verkehrssignalanlage für die Autobann.Siemens Zeitschrift 39 (1965), pp. 682-690.

/64/ Zackor,H.: Beurteilung verkehrsabhängiger Geschwindigkeitsbeschränkungen auf Autobahnen. Forschung Strassenbau und Strassenverkehrstechnik 128 (1972), pp. 1-61.

/65/ Payne, H.J., Thompson, W.A., Isaksen, L.: Design of a Traffic-Responsive Control System for a Los Angeles Freeway. IEEE Trans. on Systems, Men. and Cybernetics SMC-3 (1973), pp. 213-224.

/66/ Retzko, H.G., Cerwenka, P.: Optimierungskriterien für die Steuerung des Strassenverkehrs mit Lichtsignalanlagen. Forschung Strassenbau und Strassenverkehrstechnik 194 (1975), pp. 1-32.

/67/ Papageorgiou, M.: Ein hierarchisehes Konzept zur Regelung des Verkehrsablaufs auf Schnellstrassen. Dr.-Ing. Dissertation, Technische Universität München, 1981.

/68/ Zackor, H.: Untersuchung von Steuerungsmodellen zur Verkehrsstromführung mit Hilfe von Wechselwegweisern. Unpublished Research Report of Bundesminister für Verkehr, Feb. 1975.

/69/ Theis, T.J., Heusch, H., Boesefeldt, J.: Untersuchung der Raum-Zeit-Beziehung der werktäglichen Verkehrsstarken im städtischen und stadtnahen Bereich. Forschung Strassenbau und Strassenverkehrstechnik 218 (1976), pp. 1-109.

/70/ Wattleworth, J.A.: Peak-Period Analysis and Control of a Freeway System. Highway Res. Record 157 (1965), pp. 1-21.

/71/ Newman, L., Dunnet, A., Meis, G.J.: Freeway Ramp Control What it Can and Cannot Do. Traffic Engng. 39 (1969), pp. 14-21.

/72/ Moscowitz, K.: Motorway surveillance and control. Traffic Engng. & Control 18 (1973), pp. 526-528.

/73/ Yuan, L.S., Kreer, J.B.: Adjustment of Freeway Ramp Metering Rates to Balance Entrance Ramp Queues. Transp. Res.5 (1971), pp. 127-133.

/74/ Tabac, D.: A Linear Programming Model of Highway Traffic Control. Proc. Annual Princeton Conf. on Information Science and Systems 6 (1972), Princeton, N.J., pp. 568-570.

/75/ Wang, C.F.: On a Ramp-Flow Assignment Problem.Transp. Science 6 (1972), pp. 114-130.

/76/ Wang, J.J., May, A.D.: Computer Model for Optimal Freeway On-Ramp Control. Highway Res. Record 469 (1973), pp. 16-25.

/77/ Cheng, I.C., Cruz, J.B., Paquet, J.G.: Entrance Ramp Control for Travel Rate Maximization in Expressways. Transp. Res & (1974), pp. 503-508.

/78/ Schwartz, S.C., Tan, H.H.: Integrated Control of Freeway Entrance Ramps by Threshold Regulation. Proc. 1977 IEEE Conf.on Decision & Control, pp. 984-986.

/79/ Berkovitz, L.D.: Variational Methods in Problems of Control and Programming. J. Math. Anal. Appl. 3 (1961), pp. 145-169.

/80/ Pontryagin, L.S., Boltyanskii, V.G., Gamkrelidge, R.V., Mishchenko, E.F.: The Mathematical Theory of Optimal Processes. Interscience Publishers, 1962.

/81/ Pearson,J. B., Jr., Sridhar, R.: A Discrete Optimal Control Problem.IEEE Trans. on Automatic Control AC-11 (1966), pp. 171-174.

/82/ Halkin, H., Jordan, B.W., Polak, E., Rosen, J.B.: Theory of Optimum Discrete Time Systems. IFAC 3rd World Congress, London, June 20-25, 1966, pp. 28B.1-7.

/83/ Payne, H.J., Thompson, W.A.: Allocation of Freeway Ramp Metering Volumes to Optimize Corridor Performance IEEE Trans. Autom. Control AC-19 (1974), pp. 177-186.

/84/ Dafermos, S.C., Sparrow, F.T.: The Traffic Assignment problem for a General Network, J.Res. Not. Bureau of Standards 73B (1969), pp. 91-118.

/85/ Dafermos, S.C.: An Extended Traffic Assigment Model with Applications to two-way Traffic. Transp. Science 1971, pp. 366-389.

/86/ Branstow, W.G.: Progress in dynamic traffic assignment algorithms. Proc. Intern. Conf. on Cybern. and Society, San Francisco, 1975, pp. 369-374.

/87/ Houpt, P.K., Athans, M.: Dynamic stochastic control for freeway corridor systems, vol. I: Summary.M.I.T. Electron. Syst. Lab., Cambridge, MA, Rep. ESL-R-608, Aug. 1975.

/88/ Nguyen, S.: An Algorithm for the Traffic Assignment Problem. Transp. Science 8 (1974), pp. 203-216.

/89/ Mesarovic, M.D., Macko, D., Takahara, Y.: Theory of Hierarchical, Multilevel Systems. Academic Press, 1970.

/90/ Pearson, J.D.: Dynamic Decomposition Techniques. In "Optimization Methods for Large-Scale Systems", Wismer, D.A., Ed., McGraw-Hill, 1971, pp. 121-190.

/91/ Smith, N.J., Sage, A.P.: An Introduction to Hierarchical Systems Theory. Comput. & Electr. Engng. 1 (1973), pp. 55-71.

/92/ Singh, M.G., Titli, A.: Systems Decomposition, Optimization and Control. Pergamon Press, 1978.

/93/ Mahmoud, M.S.: Multilevel Systems Control and Applications: A Survey. IEEE Trans. Systems, Men, and Cybernetics SMC-7 (1977), pp. 125-143.

/94/ Papageorgiou, M., Schmidt, G.: On the hierarchical solution of nonlinear optimal control problems. Large Scale Systems 1 (1980), pp. 265-271.

/95/ Schmidt, G., Sendler, W.: A Failure Tolerant Multi-Microcomputer Controller for Process Control Applications, Process Automation 2 (1980), pp. 77-84.

/96/ Lasdon, L.S.: Duality and Decomposition in Mathematical Programming.IEEE Trans. Systems Science and cybernetics SSC-4 (1968), pp. 86-100.

/97/ Papageorgiou, M.: Optimal Control of Freeway Traffic by Decomposition Methods. IEEE Intern. Conf. on Control and its Applications, Warwick , 23-25 March, 1981, pp. 369-373.

/98/ Papageorgiou, M., Mayr, R.: Optimal decomposition methods applied to motorway traffic control. Intern. J. Control 35 (1982), pp. 269-280.

/99/ Singh, M.G., Tamura, H.: Modelling and hierarchical optimization for oversaturated urban road traffic networks. Intern. J. Control 20 (1974), pp. 913-934.

/100/ Schoeffler, J.D.: Static multilevel systems. In "Optimization methods for Large-Scale Systems", Wismer, D.A., Ed., McGraw-Hill, 1971, pp. 1-46.

/101/ Papageorgiou, M., Schmidt, G.: Implementation of a Hierarchical Optimization Algorithm on a Multi-Microcomputer system. IEEE Trans. on Systems, Men, and Cybernetics, to appear.

/102/ -- : Round table discussion "Trends and needs in multilevel system theory". 1st Workshop on hierarchical Control, W. Findeisen, Ed., Warsaw, 1975, pp. 305-308.

/103/ Wilson, I.D.: Foundations of hierarchical control. Intern. J. Control 29 (1979), pp. 899-933.

/104/ Lefkowitz, I.: Multilevel Approach applied to Control Systems Design. Trans. ASME J. of Basic Eng., Series D 88 (1966), pp. 392-398.

/105/ Findeisen, W. and Lefkowitz, I.: Design and Applications of Multilayer Control. 4th IFAC Congress, Warsaw, 1969, pp. 3-22.

/106/ Findeisen, W., Bailey, F.N., Brdys, M., Malinowski, K., Tatjewski, P., Wozniak, A.: Control and Coordination in hierarchical Systems. J. Willey, London, 1980.

/107/ Morari, M., Arkun, Y. and Stephanopoulos, G.: Studies in the Synthesis of Control Structures for Chemical Processes, Parts I, II. AIChE Journal 26 (1980), pp. 220-246.

/108/ Kiparissides, C., Ponnuswamy, S.R.: Multilevel control of a chain of continuous polymerization reactors. Proc. 1980 Joint Autom. Control Conf., Aug. 13-15, San Francisco, Paper TA5-D.

/109/ Labadie, J.W., Grigg, N.S., Bradford, B.H.: Automatic Control of Large-Scale Combined Sewer Systems. Proc. ASCE J. of the Environmental Engng. Div. 101 (1975), pp. 27-39.

/110/ Athans, M.: The Role and Use of the Stochastic Linear-Quadratic-Gaussian Problem in Control System Design. IEEE Trans. on Autom. Control AC-16 (1971), pp. 529-552.

/111/ Aoki, M.: Aggregation. In "Optimization Methods for Large Scale Systems - with Applications", Wismer, D.A., Ed., McGraw Hill, 1971, pp. 191-232.

/112/ Kokotovic, P.V., O'Malley, R.E., Jr., Sannuti, P.: Singular perturbations and Order Reduction in Control Theory - An Overview. Automatica 12 (1976), pp. 123-132.

/113/ Sandell, N.R., Varaiya, P., Athans, M. and Safonov, M.G.: Survey of Decentralized Control Methods for Large Scale Systems. IEEE Trans. Autom. Control AC-23 (1978), pp. 108-128.

/114/ Ikeda, M. and Siljak, D.D.: Overlapping Decompositions, Expansions and Contractions of Dynamic Systems. Large Scale Systems 1 (1980), pp. 29-38.

/115/ Saridis, G.N.: Identification Methods for Intelligent Control. Preprints of the 5th IFAC Symp. on Identifi cation and System Parameter Estimation, Darmstadt, Sept. 1979, pp. 145-149.

/116/ Irmscher, S., Reinisch, K., Thümmler, M. and Schramm, M.: Ein- und Mehrebenen Methoden zur Ermittlung von Stra tegien für die Talsperrenbewirtschaftung. 24. Intern. Wiss. Kolloquium TH Ilmenau, 1979, H. 1, A1, pp. 17-26.

/117/ Papageorgiou, M.: Mehrschichtenregelung grosser nicht-linearer dynamischer Systeme. Regelungstechnik 30 (1982), pp. 87-96.

/118/ Mehra, R.K. and Davis, R.E.: A Generalized Gradient Method for Optimal Control Problems with Inequality Constraints and Singular Arcs. Proc. Joint Autom. Control Conf. 1971, pp. 144-151.

/119/ Isaksen, L.: Suboptimal Control of Large Scale Systems with Application to Freeway Traffic Regulation. PhD Dissertation, Univ. of South. California, 1971, 230 p.

/120/ Isaksen, L., Payne, H.J.: Suboptimal Control of Linear Systems by Augmentation with Application to Freeway Traffic Regulation. IEEE Trans. on Automatic Control Ac-18 (1973), pp. 210-219.

/121/ Fu, K.S.: Sequential Methods in Pattern Recognition and Machine Learning. Academic Press, 1968.

/122/ Behrend, J., Kloss, H.: Stauuntersuchungen als Beitrag zur Verkehrsplannung and - lenkung. Strasse und Auto bahn (1970), pp. 269-274.

/123/ Breitenstein, J., Heidemann, D., Keller, H., Leichter, K., Lenz, J.H., Schulze, W., Zackor, H.: Fahrzeugpulks und Verkehrsstau. Strassenverkehrstechnik 24 (1980), pp. 16-26.

/124/ Tafel, H.J., Kumm, W.: Entwicklung eines automatischen optimalen Verkehrsbeeinflussungssystems auf Schnellstrassen.Forschung Strassenbau und Strassenverkehrs-stechnik 241 (1977), pp. 1-145.

/125/ Looze, D.P., Houpt, P.K., Sandell, N.R., Athans, M.: On decentralized Estimation and Control with Application to Freeway Ramp Metering. IEEE Trans. Autom. Control AC-23 (1978), pp. 268-275.

/126/ Saridis, G.N., Lee, C.S.G.: On hierarchically intelligent Control and Management of Traffic Systems. In "Enginee ring Foundation Conf. on Research Direction in Compu ter Control of Urban Traffic", Levine, W.S., Lieberman, E.,Fearnsides, J.J., Editors, pp. 209-218.

/127/ Blinkin, M.Ya: Problem of optimal control of traffic flow on highways. Automation and Remote Control 37 (1976), pp. 662-667.

/128/ Willsky, A.S., Chow, E.Y., Gershwin, S.B., Greene, C.S., Kurkjian, A.L.: Dynamic Model-Based Techniques for the Detection of Incidents on Freeways. IEEE Trans. Automatic Control AC-25 (1980), pp. 347-360.

/129/ Guardabassi, G., Locatelli, A., Papageorgiou, M.: A note on the optimal control of an oversaturated intersection. Transportation Research,to appear.

/130/ Lee, C.S.G., Saridis, G.N.: Hierarchically intelligent control and management of traffic systems. Preprints 8th IFAC World Congress, Kyoto, Japan, August 24-28, 1981, pp. XVII-84-89.

/131/ Drenick, R.F., Ko, K.K.: Decentralized Control of street traffic. Proc. 1982 American Control Conf., Arlington, Virginia, June 14-16, 1982, pp. 562-566.

/132/ Robertson, D.I.: TRANSYT: Traffic Network Study Tool. Traffic Engng. and Control 2 (1969).

/133/ Hunt, P.B., Robertson, D.I., Bretherton, R.D.: The SCOOT on-line traffic signal optimization technique. Traffic Engng. and Control 15 (1982), pp. 190-192.

Lecture Notes in Control and Information Sciences

Edited by A. V. Balakrishnan and M. Thoma

Lecture Notes in Control and Information Sciences

Edited by A. V. Balakrishnan and M. Thoma

Vol. 43: Stochastic Differential Systems
Proceedings of the 2nd Bad Honnef Conference
of the SFB 72 of the DFG at the University of Bonn
June 28 – July 2, 1982
Edited by M. Kohlmann and N. Christopeit
XII, 377 pages. 1982.

Vol. 44: Analysis and Optimization of Systems
Proceedings of the Fifth International
Conference on Analysis and Optimization of Systems
Versailles. December 14–17, 1982
Edited by A. Bensoussan and J. L. Lions
XV, 987 pages, 1982

Vol. 45: M. Arató
Linear Stochastic Systems
with Constant Coefficients
A Statistical Approach
IX, 309 pages. 1982

Vol. 46: Time-Scale Modeling of Dynamic Networks
with Applications to Power Systems
Edited by J. H. Chow
X, 218 pages. 1982

Vol. 47: P. A. Ioannou, P. V. Kokotovic
Adaptive Systems with Reduced Models
V, 162 pages. 1983

Vol. 48: Yaakov Yavin
Feedback Strategies for Partially
Observable Stochastic Systems
VI, 233 pages, 1983

Vol. 49: Theory and Application of Random Fields
Proceedings of the IFIP-WG 7/1
Working Conference
held under the joint auspices of the
Indian Statistical Institute
Bangalore, India, January 1982
Edited by G. Kallianpur
VI. 290 pages. 1983

Vol. 50: M. Papageorgiou
Applications of Automatic Control Concepts
to Traffic Flow Modeling and Control
IX, 186 pages. 1983